P.N.Tung 201

THE·DILEMM

AN·UPRIGH

MAX PLANCK AND TH

OF GERMAN SCI

WITH A NEW AFTE

J. L. HEILBRO

HARVARD UNIVERSIT

CAMBRIDGE, MASSACH

LONDON, ENGLA

THE DILEMMAS OF

AN UPRIGHT MAN

DRAWING OF MAX PLANCK BY
WHEN PLANCK WAS AT THE HEIG
ON THE INSTITUTIONS OF
Copyright Burndy Library, Norwalk; co
American Institute of Phy

First Harvard University Press paperback edition, 2000

Library of Congress Cataloging-in-Publication Data
Heilbron, J. L.
The dilemmas of an upright man : Max Planck and
the fortunes of German science / J.L. Heilbron.
p.   cm.
Originally published: Berkeley : University of California Press, c1986.
With new afterword.
Includes bibliographical references and index.
ISBN 0-674-00439-6
1. Planck, Max, 1858–1947.   2. Physics—Germany—History.
3. Physicists—Germany—Biography.   I. Title.
QC16P6 H45   2000
530′.2.092—dc21
[B]   00-040745

CONTENTS

CONTENTS

PREFACE

For several years around the turn of the century, Max Planck was the dean and definition of theoretical physics in Germany. In 1900 his long and relentless battle with an obscure technical problem culminated in a revolution that he neither wanted nor welcomed: he initiated the quantum theory, which conflicts with ordinary physical ideas but forms the foundation of modern physics.

Soon after Planck made this capital contribution to the substance of science, he took on responsibility for shaping its institutions and philosophy. He collected central positions in several German agencies that supported physics and lectured widely on the nature of nature and the theories invented to comprehend it. His science policy and his philosophy of nature, his physics and his personal life, were at first all of a piece, and at one with his conceptions of self, state, and society. But even as his institutional authority expanded during the Great War, the Weimar Republic, and the Third Reich, the unity of the world picture he had developed under the Empire shattered. The consequent reconstruction of his world picture and his politics makes up most of the matter of this book.

Planck's maneuvers have a particular claim on our attention because he was a man of exceptional probity placed in positions of exceptional sensitivity. He made no decision lightly. His struggle to come to grips with changing political circumstances and his efforts to adjust the image of science and civic life that he developed before 1900 to the realities of twentieth-century Germany have the elements of heroic tragedy. His life is a lesson.

Planck's library and all his private papers were destroyed by Allied bombing in 1944. Information about him must therefore be sought in the archives of the institutions he served and in the manuscript remains of his correspondents. Much of the record of his public service and indications of his private life survive in these documents. I have profited greatly from the labors of scholars in the German Democratic Republic, especially the authors of the invaluable *Studien zur Geschichte der Akademie der Wissenschaften der DDR*. The gathering of Planck's correspondence has been facilitated by the Inventory of Sources for Twentieth-Century Physics, a creation of the Office for History of Science and Technology, University of California, Berkeley,[1] and by the kindness of librarians in Europe and the United States who responded generously to requests for photocopies. Among those to whom I am particularly indebted for access to documents are David Cassidy (Einstein Papers, Boston University), Armin Hermann (Universität Stuttgart), Dietrich Hoffmann (Akademie der Wissenschaften der DDR), Otto Mayr and Rudolf Heinrich (Deutsches Museum, Munich), Robin Rider (The Bancroft Library, Berkeley), and Silva Sandow (Max-Planck-Gesellschaft). The text has profited from critical readings by Alan Beyerchen, Gerald Feldman, Barbara Reeves, and Spencer Weart.

It is a pleasure to thank the following institutions and individuals for permission to quote correspondence in their collections: American Philosophical Society, Philadelphia; Archive for History of Quantum Physics and The Bancroft

1. This resource, which will be available soon on microfiche, is described in Wheaton, *Isis*, 75 (1984), 153–57.

Library, University of California, Berkeley; Niels Bohr Archive, Niels Bohr Institut, Copenhagen, and Prof. Aage Bohr; Bundesarchiv, Koblenz; Cornell University Library, Ithaca, N.Y.; Deutsches Museum, Munich; Einstein Papers, Hebrew University, Jerusalem, and the American Friends of the Hebrew University; Göttingen University Library, Göttingen; Heinrich-Heine-Institut, Düsseldorf; Kungl. Vetenskaps Akademien, Stockholm (Nobel Archives); Schiller-National-Museum, Marbach; the Estate of Max Planck; Max-Planck-Gesellschaft, Berlin; Museum Boerhaave, Leyden; Sächsische Landesbibliothek, Dresden; Oslo University Library, Oslo; Rensselaer Polytechnic Institute, Troy, N.Y.; Rockefeller Archive Center, North Tarrytown, N.Y.; Staatsbibliothek Preussischerkulturbesitz, Berlin; Society for Psychical Research, London; Stockholm University Library, Stockholm; Tübingen University Library, Tübingen; Zentrales Staatsarchiv, Potsdam.

This book is the unexpected outcome of a bibliography of Planck's nontechnical writings issued as a byproduct of the Inventory of Sources.[2] The bibliography prompted Hans Rotta of Hirzel-Verlag, a major publisher of Planck's works, to propose a reprinting of Planck's enduring essays. The planned introduction to the reprinting has swollen to the present book. I am most obliged to Mr. Rotta for his initiation of this project and for his forbearance as it grew. The German edition contains both a translation of the following text and a reprinting of Planck's most important essays on the nature of science. It also gives the original texts of Planck's correspondence quoted here in translation.

2. Lowood, *Max Planck* (1977); a second edition of this bibliography is in preparation.

ABBREVIATIONS

Only authors' names and short references to their works appear in the notes; full publishing information is given in the bibliography. The following abbreviations are used in both the Notes and the Bibliography:

ADB Alan D. Beyerchen. *Scientists under Hitler*. New Haven: Yale University Press, 1977.

AE   Albert Einstein Papers. Hebrew University, Jerusalem.

AH   Armin Hermann. *Max Planck in Selbstzeugnissen und Bilddokumenten*. Reinbeck: Rowohlt, 1973.

AHQP Archive for History of Quantum Physics. Office for History of Science and Technology, University of California, Berkeley.

ANP Artur Neuberg Papers. Sächsische Landesbibliothek, Dresden.

APS   American Philosophical Society, Philadelphia.

AW   Akademie der Wissenschaften.

DM   Deutsches Museum, Munich.

DPG Deutsche Physikalische Gesellschaft.

ESC	Ehrenfest Scientific Correspondence. Museum Boerhaave, Leyden, and AHQP.
GDNA	Gesellschaft Deutscher Naturforscher und Ärzte.
HH	Hans Hartmann. *Max Planck als Mensch und Denker*. 4th ed. Basel, Thun, and Düsseldorf: Ott, 1953.
HSPS	*Historical studies in the physical sciences*.
ITGOW	Institut der Theorie, Geschichte und Organisation der Wissenschaft.
JW	*Jahrbuch für Wirtschaftsgeschichte*.
KA	Anton Kippenberg Papers. Schiller-National-Museum, Marbach.
Kirsten and Körber	Christa Kirsten and Hans-Günther Körber. *Physiker über Physiker*. 2 vols. Berlin: Akademie-Verlag, 1975–79.
Kirsten and Treder	Christa Kirsten and Hans-Jürgen Treder. *Albert Einstein in Berlin, 1913–1933*. 2 vols. Berlin: Akademie-Verlag, 1979.
KWG	Kaiser-Wilhelm-Gesellschaft.
LS	Leo Stern, ed. *Die Berliner Akademie der Wissenschaften in der Zeit des Imperialismus*. 3 vols. Berlin: Akademie-Verlag, 1975–79.
LTZ	Hendrik Antoon Lorentz Papers. Museum Boerhaave, Leyden, and AHQP.
MPG	Max-Planck-Gesellschaft.
NDW	Notgemeinschaft der deutschen Wissenschaft.
Nwn	*Die Naturwissenschaften*.

PA   Max Planck. *Physikalische Abhandlungen und Vorträge*. 3 vols. Braunschweig: Vieweg, 1958.

PB   *Physikalische Blätter*.

PhW   Heinrich Vogel. *Zum philosophischen Wirken Max Plancks. Seine Kritik am Positivismus*. Berlin: Akademie-Verlag, 1961.

PPA   Akademie der Wissenschaften, Berlin. *Max Planck in seinen Akademie-Ansprachen*. Berlin: Akademie-Verlag, 1948.

RF   Rockefeller Foundation Archives, North Tarrytown, New York.

RP   Carl Runge Papers. SPK.

Sb   AW, Berlin. *Sitzungsberichte*.

SPK   Staatsbibliothek Preussischerkulturbesitz, Berlin.

TBL   The Bancroft Library, University of California, Berkeley.

VWPS   *Vierteljahrschrift für wissenschaftliche Philosophie und Soziologie*.

Wege   Max Planck. *Wege zur physikalischen Erkenntnis*. 3rd ed. Leipzig: Hirzel, 1944.

ZPCU   *Zeitschrift für den physikalischen und chemischen Unterricht*.

ESTABLISHING THE
WORLD PICTURE

Max Planck, exemplar of the classical physicist, came from a line of pastors, scholars, and jurists. To go back no further than three generations, his great-grandfather Gottlieb Jakob Planck (1751–1833), a student of immediate disciples of Leibniz, was professor of theology for almost fifty years at the University of Göttingen. His Enlightenment ideals of rationalism and tolerance, reexpressed in late-nineteenth-century terms with God, not Christ, at the center of belief, grounded his great-grandson's firm, liberal, ecumenical religion.

Gottlieb Planck's example descended through his son, likewise a Göttingen theologian, and his grandson, Wilhelm Johann Julius Planck, who reversed the emphasis in the family's fascination with church law and became a professor of jurisprudence. Wilhelm Planck taught until 1867 at Kiel, and then in Munich, where he died in 1900. His sixth child, Max, was born in 1858. Wilhelm's second wife, Max Planck's mother, came from a family of pastors. She is said to have been a lively, even passionate, woman. Planck remained attached to her until her death, at the age of ninety-three, in August 1914.[1]

1. HH, 29–40; Dinkler, *Zeitschrift für Theologie und Kirche*, 56 (1959), 215, 221; *Realencyclopädie für protestantische Theologie und*

Planck's strong ties to his family extended to his aunts and cousins, with whom as a boy he shared the subdued amusements of the children of Wilhelmian professors, lawyers, clerics, and high government officials. A few glimpses of this life can be caught through letters: summers at the Baltic resort of Eldena, which he remembered as the paradise of his childhood; croquet on the lawn; evening readings from the novels of Sir Walter Scott; skeet shooting; plays; musicales.[2] We see Planck as an earnest young man exchanging impressions of life with equally earnest friends, who wrote out their thoughts for two weeks in turn in a large book that has not survived. There were trips to Austria and Italy, a courtship that won a banker's daughter, Marie Merk, and excursions to an uncle's hunting lodge in East Prussia.[3]

In 1885, on one of these excursions, Planck met a man six years his junior, a student of physics whose experimental and theoretical work was to be the starting point of his own most important contribution to science. The young man was Wilhelm Wien, son of a Prussian landholder, who became and remained a friend and collaborator of Planck's for over forty years. Their long friendship and shared professional interests did not extend to political alliance. The Plancks, although conservative on an ordinary scale, may appear almost liberal in comparison with the reactionary and chauvinistic Wiens, who thought the firing of Bismarck the greatest disaster in German history and regretted that the emperor lacked the power to destroy his grandmother Victoria's nation of shopkeepers. Planck's father did not ap-

---

Kirche, s.vv. "G. K. Planck" and "H. L. Planck." Cf. Born, Royal Society of London, *Obituary notices*, 6 (1948), 161–88, and Unsöld, *Physik* (1958), 9.

2. Planck to Carl Runge, 11 Oct. 1877 (RP); to "Emmachen" (Frau Max Lenz, his first cousin), 21 Nov. 1917; to Hildegard Gravemann (niece), 27 Feb. 1944; and to Dora Martin (cousin), 23 May 1942 (MPG).

3. Iris Runge, *Carl Runge* (1949), 34; Planck to Carl Runge, 31 July 1877 and 24 Jan. 1880 (RP).

prove Bismarck's policy of annexation, and neither father nor son would have countenanced a preventive war against England.[4] And Planck regarded himself as more liberal politically and socially than most of his family.[5] But he was no more a social than a scientific revolutionary. A conservative in the root meaning, his particular effectiveness lay in his ability to adapt to, and even direct, current realities while saving, and acting on, traditional values.

Planck was not a genius. His teachers at the Maximilians-Gymnasium in Munich ranked him near, but never at, the top of his class: fourth of twenty-eight in 1868–69, then fifth of thirty-seven, eighth of twenty-three, third of twenty-one, fourth of nineteen. He did well in everything—languages, mathematics, history, music—and was extremely diligent and dutiful, but his teachers noted no special brilliance or aptitude, except in his personal qualities. They esteemed the quiet force of his personality, his shy strength of character, and they judged him to be, "deservedly, the favorite of his teachers and classmates."[6]

Planck's success in physics, for which he always thought he had no particular gift,[7] and in his many other intellectual and administrative activities came from a long absorption in the material and a slow maturing of ideas. He did not run after novelties ("for by nature I am peaceful and disinclined to questionable adventures") or respond to them spontaneously ("for unfortunately I have not been given the capacity to react quickly to intellectual stimulation").[8] He ex-

4. Planck, in W. Wien, *Aus dem Leben* (1930), 139; Wien, ibid., 29–30, 62; Planck, "Selbstdarstellung" (1942), 5.
5. Planck to Frieda Clarke (niece), 2 May 1925 (APS).
6. Information from copies of the records of the Köngl. Maximilians-Gymnasium, kindly supplied by Armin Hermann.
7. Planck to Joseph Strauss, 14 Dec. 1930, in Balmer, *PB, 25* (1969), 558.
8. Respectively, Planck to R. W. Wood, 7 Oct. 1931, in Hermann, *Frühgeschichte* (1969), 31, and to Bohr, 7 May 1920, in Bohr, *Works, 3,*

pressed astonishment at the ability of others to pursue several research lines simultaneously; as he wrote Arnold Sommerfeld, a contemporary theoretical physicist and long-time friend, he found it very difficult "to leave a subject quickly after I've worked my way into it and to take it up [again] quickly at a favorable opportunity."[9] "Quickly" was not his speed. But once Planck had mastered something, he understood it with that force and clarity of intellect that, according to Descartes, are our best guarantee of the truth of our opinions.

Planck's confidence in himself and his ideas increased in step with Prussia's triumphs on the battlefield and with the new Reich's rise to dominance among the European nations. Although personally the most modest of men, Planck identified his own development so fully with Germany's that the preservation of its cultural capital was inseparable from the preservation of his personal values and professional life. Over all these values stood the ideal of unity, which in the political sphere inspired the creation of the Wilhelmian empire and in the cultural sphere inspired belief in the interconnectedness of all respectable branches of learning. Planck's pride in imperial Germany and his commitment to the academic ideal of the unity of knowledge were the pillars on which he raised his science policy.

Respect for law, trust in established institutions, observance of duty, and absolute honesty—indeed sometimes an excess of scruples—were the hallmarks of Planck's character. His modesty in the face of his virtues, his renown, and his authority were frequently remarked; his contemporaries respected the man as much as they admired the scientist. To

---

677. Cf. Planck to Runge, 9 Dec. 1878 (RP), complaining about insufficient time to finish an easy examination.

9. Respectively, Planck to Albert Schweitzer, 20 Dec. 1930, in Kangro, *Vorgeschichte* (1970), 123, and Planck to Sommerfeld, 1 July 1923, in AH, 68; cf. Sommerfeld, *Nwn*, 5 (1918), 199.

take but one example, when his longtime colleagues at the
Berlin Academy observed the fiftieth anniversary of his doctorate in 1929, they praised not only his science but also,
what is less common among academics, "the spotless purity
of his conscience."[10] Planck valued a clear conscience as the
greatest blessing a human being can enjoy.

The inner grace of clarity of conscience may appear to the
outside world as a great impediment to change of mind. One
of Planck's students, Walther Meissner, had the impression
that Planck never was "swayed by the opinions of others, not
only in science but also in human relations." His clear conscience was the only compass he needed. Planck once described to his student and successor, Max von Laue, his way
of fixing his direction: "My maxim is always this: consider
every step carefully in advance, but then, if you believe you
can take responsibility for it, let nothing stop you."[11] Planck
of course did change his opinions, and on important matters,
but he did not change them easily.

It was not his stubbornness or his probity that made
Planck the chief spokesman for German science between the
world wars, although he needed both, and much patience
too, in his dealings with colleagues and bureaucrats. Planck's
domestic power derived from his reputation as a German
natural philosopher who had changed the course of international science.

Planck's Formula

Planck's upsetting of physics, which dates from the turn of
the century, resulted from a compromise on his part, and

10. *PPA*, 127 (1929). On Planck's modesty, see the contributions by A.
Bertholet, H. Ficker, and E. Lamla to "Max Planck zum Gedächtnis,"
PB, *4* (1948), 162, 174; AH, 76.

11. Meissner, *Science*, *113* (1951), 75; Planck to Laue, 22 Mar. 1934,
in AH, 86.

even a capitulation. It followed the uncomfortable realization that he could not solve a problem on which he had worked for several years in the terms in which he had conceived it. The problem might seem scarcely worth the attention, much less the discomfort, of a grown man. Planck had set himself the task of calculating from first principles the relative intensity of each color in the radiation emerging from a small hole in a special electric oven or cavity. Very loosely put, he wanted to be able to compute the tints ("red hot," "white hot," etc.) of the inside of a furnace for any assigned temperature.

There were two dissimilar reasons for desiring to know the color intensities, or energy distribution, of the cavity radiation. Fundamental and plausible arguments showed that at equilibrium the distribution does not depend on the size or shape of the cavity or on the material of its walls. A formula for the distribution would provide, for a unit volume of the cavity, a specification of the energy carried by the light waves of each color. The formula would consequently contain only the temperature, the length of the waves (a measure of color), and one or more "universal constants," the same for all cavities, colors, and distributions. For Planck, finding the formula meant discovering a relationship of very general theoretical interest, something "independent of special bodies and substances, which will necessarily retain its importance for all times and cultures, even for nonterrestrial and nonhuman ones."[12]

A second reason for concern with cavity radiation was entirely practical. The equilibrium distribution maximizes the amount of heat (longer wavelengths) for a given amount of light (shorter wavelengths). Hence the formula Planck sought described the worst possible source of illumination, and so could serve as a zero-point standard for rating new

12. *PA, 1,* 599–600 (1899).

electric lamps. Accordingly, the Physikalisch-Technische Reichsanstalt, the imperial bureau of standards, took an interest in measuring the equilibrium distribution of cavity radiation. In 1896 Wilhelm Wien and others in the Reichsanstalt's physics department put together an expensive empty cylinder of porcelain and platinum and recorded the color distribution of radiation allowed to escape from a hole in one of its ends. They worked primarily with shorter wavelengths, from the near infrared into the violet. At the Berlin Technische Hochschule, another of Planck's close associates, Heinrich Rubens, operated another oven and pursued the measurements into the deep infrared. A good many empirical formulas were concocted, which more or less fit the facts. For a time a semitheoretical formula proposed by Wien in 1896 seemed the best.[13] Planck set out to derive Wien's formula from the basic laws of electro- and thermodynamics.

Because the equilibrium distribution does not depend upon the nature of the walls, Planck was free to model them in the most convenient manner. He represented cavity walls as a collection of "resonators" and each resonator as a massless spring with an electric charge at its end. The springs come in all possible stiffnesses and therefore can oscillate at all possible frequencies. Heating the walls puts the springs in motion; the accelerating charges radiate energy into the cavity in accordance with Maxwell's electrodynamics and also absorb colors with whose waves they resonate; equilibrium eventually sets up, in which all the resonators of a given frequency absorb as much energy from the cavity radiation as they put into it.

Maxwell's equations tell, or rather then told, everything about the emission, absorption, and propagation of radiation from the resonators, but they said nothing about the energy

13. Kangro, *Vorgeschichte* (1970), secs. 5.4, 7.2.

distribution at the ultimate equilibrium. Planck thought that by making a special assumption relating the average energy of the resonators to their entropy he could obtain Wien's formula. (The physical significance of entropy, a favorite concept of Planck's, will occupy us presently.) His optimism was rudely shattered by criticism from the world's authority on such problems, Ludwig Boltzmann, who had studied in detail the equilibrium and the approach to equilibrium of gases. Although Planck had not approved Boltzmann's gas theory, he admired its author;[14] he accepted the criticism Boltzmann drew from his theory, found another special assumption about entropy and energy, and rederived Wien's formula. That was in 1899. Since the formula still fit the facts, the job seemed done, except for the annoyance that Planck's special assumption could not be firmly grounded or others ruled out.[15]

In 1900 experimenters at the Reichsanstalt, by then pushing into the infrared, found irreducible discrepancies between Wien's law and their measurements. In March of that year Planck presented to the Berlin Academy an improved plausibility argument in favor of his special assumption; but in October, bending under the weight of the facts, he reduced Wien's law to an approximation valid only at short wavelengths. Planck's conscience would not allow him to believe for long in theories uncontestably in disagreement with exact measurement. That would be to act the philosopher. As he wrote a few years later to Wien, criticizing the British physicist James Jeans for resisting evidence stronger than the discrepancies that had convinced Planck, "He is the model of a theorist as he should *not* be, just as Hegel was in philosophy: so much the worse for the facts if they don't fit."[16]

14. Planck to L. Koenigsberger, 7 Feb. 1895 (acc. Darmst., 1922.93, SPK).

15. Kuhn, *Black-body theory* (1978), 6–11, 82–91, 278.

16. Planck to Wien, 27 Feb. 1909 (AHQP); cf. *PA*, *2*, 247 (1910).

Theoretical physicists, like nature, abhor vacuums: having discarded Wien's formula as truth, Planck put in its place another—a result of inspired guesswork, scientific tact, sober compromise, in short, of tinkering. The new relation agreed with all reliable measurements, and it did so with two empirical universal constants, no more than Wien had required.

So far the compromise. Next came the capitulation. In trying to derive his compromise formula, Planck recast it to express the entropy of all the oscillators at a given frequency. Here he discovered a strong analogy to Boltzmann's formulation of the entropy of a gas at equilibrium. Boltzmann's formulation rested on the proposition that equilibrium is not a final and unchanging state but only the most likely condition. Planck had been, and for a time remained, most reluctant to accept this proposition. It menaced his conception of science and of his own life's work.

Reluctant Atomist

Planck drew the inspiration for his life's work from the grand synthesis of classical physics effected by Rudolf Clausius and William Thomson in the 1850s: thermodynamics, or the doctrine of the conservation, transformation, and degradation of energy. This doctrine spoke to the lawyer and the theologian in Planck. It claimed a universal dominion to which all the laws of physics had to conform, and it dealt with the permanent and the unchanging, which physicists since Aristotle had identified with the true and the good. Such lofty considerations, Planck wrote at the end of his long life, had brought him to physics instead of to mathematics or history or music.[17] The choice of physics did not then confer the sort of prestige on the chooser that it does

17. Planck, *Erinnerungen* (1948), cols. 116–19; *PA, 3*, 374 (1948); cf. AH, 7–13.

today. The humanities dominated the faculties. "At that time educated people generally considered the natural sciences to be an inferior sort of knowledge. I remember well [Planck told his first biographer] that my cousin and colleague [at the University], the historian Max Lenz, would refer jokingly to us scientists [*Naturforscher*] as foresters [*Naturförstern*]."[18] Planck had not worried about worldly success. "I have rather worked steadily from an inner compulsion."[19]

Planck traced the discovery of his vocation to the teaching of an instructor at the gymnasium, Hermann Müller, who awakened an interest, which became a passion, to "investigate the harmony that reigns between the strictness of mathematics and the multitude of natural laws." In 1878, at the age of twenty, Planck chose thermodynamics as the subject of his doctoral dissertation, which he wrote in four months. He recalled that his professor at the University of Munich, Philipp von Jolly, had counseled against a career in physics on the ground that the discovery of the principles of thermodynamics had completed the structure of theoretical physics. That had not dissuaded Planck, who had his compulsion and also an objective far removed from the principal ambition of today's physicists. He had no wish to make discoveries, he told Jolly, but only to understand and perhaps to deepen the foundations already set.[20]

Planck's doctoral dissertation of 1879 reviewed the two principles of classical thermodynamics. The first states the conservation of energy; the second, in what Planck called its most general significance, establishes a direction in time by defining a quantity, entropy, that increases in all real physi-

18. Planck to Hans Hartmann, 4 July 1942 (MPG).

19. Balmer, *PB, 25* (1969), 558.

20. Planck to Runge, 9 Dec. 1878 and 4 Mar. 1879 (RP), reporting beginning and ending his thesis; Planck, "Selbstdarstellung" (1942), 4, 6.

cal processes.[21] In his first original scientific work, his *Habilitationsshrift* of 1880, devoted to the "equilibrium states of isotropic bodies," Planck extracted some concrete and testable results from the principles of thermodynamics without making assumptions about the internal structure of substances. He added only the premise that stable equilibrium comes at maximum entropy, after which, according to his strict construction, no further change in the system can occur. Planck laid great stress on the capacity of thermodynamics to produce useful results without reliance on the atomic hypothesis, which (so he wrote in 1882) he regarded as, at best, an auxiliary to be invoked only after deductions from the laws of energy and entropy had been exhausted.[22]

Planck's praise of the entropy principle and his few proposals for experiments made little impression on physicists. His professors at Munich did not understand his dissertation; in Berlin, where he had studied for a year, Gustav Kirchhoff, the professor of theoretical physics, judged it wrong, and Hermann von Helmholtz, a codiscoverer of the energy principle, judged it not at all; as for Clausius in Bonn, Planck could not catch him to serve him with a copy.[23] Whether read or not, his work justified the University of Kiel, where his father still had friends and perhaps influence, to engage Planck as "extraordinary" (associate) professor of theoretical physics in 1885. There were then very few posts in the very new field of theoretical physics. "Needless to say," Planck wrote his friend Carl Runge, "I am very lucky."[24]

Soon after taking up his new post, Planck pounced on an opportunity to correct what he thought to be an injustice to

21. *PA*, *1*, 1, 3 (1879); cf. ibid., 197 (1887), 427 (1892).

22. Ibid., 62–63 (1880), 140, 161 (1882); cf. ibid., 197, 202 (1887).

23. Planck, *Erinnerungen* (1948), cols. 121–26; *PA*, *3*, 377–79 (1948); AH, 14–20.

24. Planck to Runge, 13 Apr. 1885 (RP).

himself and to thermodynamics. He entered a prize competition on the "nature of energy" offered by the Philosophical Faculty of the University of Göttingen. Planck won second prize, no first being given; he inferred from the judges' criticism of his work that his offhand treatment of Göttingen's professor of physics, Wilhelm Weber, had caused the Göttingen faculty to award him the lower prize.[25] But in this there was compensation. Quasi-rejection by Göttingen assisted quasi-acceptance by Berlin.[26] In 1889 Planck was called there, still as associate professor, to succeed Kirchhoff.

He found the capital still resistant to the gospel of thermodynamics and ignorant of the requirements of the entropy principle. His first performance at the Deutsche Physikalische Gesellschaft (the German Physical Society), of which he later became an ornament, was not a success. Planck filled the blackboard with his latest results on electrolytes, which accounted nicely for some recent experiments, and finished his lecture with the news that he had probably made a decisive contribution to a fundamental problem. No one said a word. Then the chairman, Emil du Bois-Reymond, feeling obliged to break the silence, spoke critically. Planck recalled, "That was quite a shower on my glowing enthusiasm. I returned home somewhat depressed but soon consoled myself with the thought that a good theory will succeed without clever propaganda."[27] Planck's views did ultimately prevail; by 1891 he had had to lend out his thesis so often to colleagues who wished enlightenment

25. Planck, *Erhaltung* (1908), xii–xiii (text of 1887). The faculty praised "the author's methodical thought, his thorough mathematical-physical training, [and] the discretion of his judgment" (ibid., x).

26. Planck, *Erinnerungen* (1948), cols. 127–30; *PA, 3*, 380–81 (1948).

27. *PA, 1*, 371 (1890); Planck, *Erinnerungen* (1948), cols. 137–40; cf. AH, 23.

about entropy that it was ready to fall apart.[28] No doubt the happy experience of being almost right in the end strengthened his native confidence and resolve.

In 1892 Planck rose to full professor. Two years later he became an ordinary (full) member of the Berlin Academy of Sciences on the nomination of Helmholtz, who praised Planck for securing by thermodynamics certain results that physical chemists had obtained from special hypotheses about atoms and ions. Planck felt pleased and even vindicated by Helmholtz's prediction that from his point of view previously disjointed parts of physics and chemistry would appear as a whole.[29] It seemed that, undeterred by the indifference and doubts of others, Planck had marched unerringly to the summit of science. On pulling his various investigations together into a uniform text in 1897, he wrote that he had found nothing of importance that needed changing. He had become and remained the world's authority on classical thermodynamics.[30] As we know, he did not change his mind easily.

An essential alteration would be required, however, to extend thermodynamics outside the range of thermochemical problems with which Planck had been engaged. This alteration overturned Planck's conception of the foundations of

28. Planck to Ostwald, 16 July 1891, in Körber, ed., *Briefwechsel Ostwalds* (1961), *1*, 35.

29. Helmholtz, "Wahlvorschlag für Max Planck," in Kirsten and Körber, *1*, 125–26; Einstein, *Nwn, 1* (1913), 1077. Cf. Hans Hartmann, *Naturwissenschaftliche Rundschau, 11* (1958), 128–31, and Born, Royal Society of London, *Obituary notices, 6* (1948), 163–66, on Planck's several complaints about delayed recognition.

30. Planck, *Thermodynamik* (1905), "Vorwort," iii (text of 1897). Cf. his correspondence with Wilhelm Ostwald between 1891 and 1898, in Körber, *Briefwechsel Ostwalds* (1961), *1*, 51–67; Planck to Mach, 25 June 1893, in Heller, *Ernst Mach* (1964), 28–29; Planck to Alfred Landé, 3 Feb. 1926 (AHQP 4/19).

the subject: it amounted to restricting the sweep of the entropy principle and riddling the law with loopholes. He began to acquiesce in this act of lèse-majesté in 1900, in order to secure his radiation law. It was the penultimate step in an internal debate that had lasted for two decades about the relation between the entropy principle and the atomic hypothesis.

As a student in Munich Planck had heard the fulminations of the chemist Hermann Kolbe against atomism as a gratuitous and misleading interpretation of chemical regularities. By 1881 he had learned for himself that atomism might not conduce to the progress of science. It then appeared to him that the atomic conception of matter, which mandates no direction for natural processes, stood in irreconcilable opposition to the principle of increase of entropy. He expected that there would be a fine and fatal fight between the two theories, and he had a good idea which would lose: "Despite the great success that the atomic theory has so far enjoyed, ultimately it will have to be abandoned in favor of the assumption of continuous matter."[31]

Although Planck was to sustain for almost twenty years the hope that continuum mechanics might be compatible with the entropy principle, he soon found that he could not stay at the forefront of thermochemistry without recourse to a molecular view of matter. Equipped only with the principles of thermodynamics, he could not have developed and helped to propagate the new physical chemistry of Svante Arrhenius and J. H. van't Hoff, which occupied him from 1887 to the mid-1890s. As he told a correspondent in 1890, physicists had no alternative but to descend into the molecular world for explanations of the conductivity of liquids and osmotic pressure, for no other fruitful theory existed.[32] In

31. *PA*, *1*, 163 (1882); ibid., *3*, 148–49 (1924).
32. Planck to an unnamed colleague, 8 July 1890 (DM).

1887 Planck included Avogadro's hypothesis about the molecular constitution of gases among "those laws . . . that seem to us the most certain foundation of theoretical inquiry." He wrote without qualification about the numbers of molecules in solutions and about the behavior of ions there. In his Göttingen prize essay he pointed to the natural association of the principle of energy conservation with the widespread belief that all natural processes might be represented as mechanical, and he explicitly distanced himself from doubters who claimed that the program of mechanical reduction rested on "entirely too narrow a conception of natural phenomena."[33]

Sed contra, the most striking success in mechanical reduction since the establishment of thermodynamics in the 1850s had been achieved by the kinetic theory of gases, and just there, Planck told the annual meeting of German scientists and physicians (the Deutsche Naturforscherversammlung) in 1891, progress no longer seemed proportional to effort. Anyone familiar with the "awesome exercise of physical acuteness and mathematical virtuosity" that had been devoted to following molecular motions could scarcely help concluding, he said, that the game was not worth the candle.[34] Hence it behooved the analyst to get on without the atomic hypothesis, or indeed any other hypothesis, as far as he could: earlier hypotheses, however successful, became a present danger, "the most dangerous enemy of progress," once their work was done.[35] In any case, as Planck told the

33. *PA, 1,* 271, 275 (1887), 289, 294–95 (1889), 350 ff. (1890); Ostwald, *Lebenslinien* (1926), *2,* 30–31; Planck, *Erhaltung* (1908), 1, 58 (1887).

34. *PA, 1,* 372–73 (1891). Cf. Planck to Ostwald, 25 Sept. 1891, in Körber, ed., *Briefwechsel Ostwalds* (1961), *1,* 38, and Ostwald, *Lebenslinien* (1926), *2,* 187–88.

35. Planck, *Thermochemie* (1893), iii–iv. The same message appears in Planck to Ostwald, 1 July 1893, in Körber, ed., *Briefwechsel Ostwalds* (1961), *1,* 50.

Berlin Academy in 1894 on the occasion of his entry into it, mechanical reduction, whether to atoms or to anything else, could scarcely appeal to a tasteful theorist, since it was insufficiently prescriptive: contrary to the standard objection that mechanics provided too narrow a basis for representing physical phenomena, Planck deprecated it as too diffuse: "There is now no single, definite mechanical model, but rather a number of entirely different ones, each of which claims to reflect the course of individual physical processes. . . . All are extremely complicated, and none has decisive advantages over the others."[36]

But yet again, the new physical chemistry had prospered by applying to solute molecules the considerations and equations of a rudimentary kinetic gas theory. As Planck and others who nominated van't Hoff for a unique professorship at the Berlin Academy trumpeted, their man was "the creator of a fruitful new branch of molecular physics."[37] Were van't Hoff's insight developed in accordance with the laws of thermodynamics, Planck had told the Naturforscherversammlung, it might vouchsafe a knowledge "that allows a deeper look into the world of molecules than even the kinetic theory itself, which explicitly starts from definite ideas about this world."[38]

Warming now to the indications of the ionic theory of solutions, Planck accepted the possibility and endorsed the desirability of a definitive mechanical reduction. Only mechanics, strengthened by the energy principle, seemed to offer the basis for unification: "The deepest form of coherence . . . lies in identity, and this cannot be achieved in any physical field better than in mechanics."[39] In this mood Planck could consider thermodynamic analysis merely a pis aller,

36. *PPA*, 2 (1894); Kirsten and Körber, *2*, 169–70 (1894).
37. Kirsten and Körber, *1*, 135–36 (1895); Planck to Emil Fischer, 13 Jan. 1895 (TBL).
38. *PA*, *1*, 373 (1891).
39. *PPA*, 1, 4 (1894); Kirsten and Körber, *1*, 169, 170 (1894).

appropriate not to the power but to the ignorance of science. It might make better sense to tie together the various threads of physics provisionally through the concepts of energy and entropy than to aspire all at once to weave the final tapestry of the mechanical world picture. The weaving could not be postponed indefinitely, however: the nature of science required, and the first law of thermodynamics encouraged, the theorist to take to his loom.[40]

CONVERSION

The tensions, not to say contradictions, between these various pronouncements were heightened in 1895 by Planck's assistant Ernst Zermelo, who appeared to have closed a loophole permitted by the usual argument against a mechanical representation of entropy. An example might help. We begin with two equal volumes of the same gas, one hot, the other cold, separated by a wall impenetrable to heat and molecules. The wall now falls and the gases mix, coming to equilibrium at an intermediate temperature. If at equilibrium the velocities of all the molecules were miraculously reversed, the gas would (or might, since the process had never been calculated) return in due course to its initial state, in contradiction to the entropy principle. Nothing in mechanics prevents the existence of the velocity-reversed state and therefore of a process that decreases entropy. Physicists had evaded the logical end of this argument by declaring that velocity-reversed states, although allowable, do not occur. According to Zermelo's new argument, however, the molecules of a gas will spontaneously assume velocity-reversed states because, given enough time, the laws of motion will return them to, or very close to, their initial positions and velocities. If their entropy increases at some part of the cycle, it must decrease in another.[41]

40. *PPA*, 4 (1894); Kirsten and Körber, *1*, 170 (1894); Planck, *Thermodynamik* (1905), v (1897).

41. Dugas, *Théorie physique* (1959), 192–208.

Boltzmann replied to this arresting objection by repeating his interpretation of the second law of thermodynamics as a statistical principle: entropy does not always increase, just almost always. To defeat Zermelo's argument, Boltzmann had to say that the amount of time spent by a gas in equilibrium in moving from states of higher to states of lower entropy is vanishingly small in comparison with the time spent in moving in the opposite direction. The physical reason for this huge disparity, according to Boltzmann, is that the number of molecular configurations associated with increasing entropy is overwhelmingly larger than the number associated with decreasing entropy.[42] Planck supported Zermelo in the discussion with Boltzmann, although not to the point of ruling out the possibility of a mechanical underpinning of the entropy principle.

In the spring of 1897, in a most interesting letter to his former fellow student Leo Graetz, Planck outlined a compromise position.[43] If the molecules in a particular sample of gas happen to be in an improbable configuration, the gas will develop in an odd way, perhaps separating into colder and warmer parts, merely in accordance with the laws of mechanics: once the initial configuration or state is specified, probability no longer plays a part. According to Planck, who still would not let entropy decrease, Boltzmann had to rule out all initial states whose development brought any backsliding from advancing entropy. For that there seemed no justification: "It would be completely unjustified to assume that changes in nature always take place from less to greater probability." From this it appears that Planck did not hold Boltzmann's approach to be absolutely wrong but—to use a phrase from the gas theory—only infinitely improbable.

It was certainly easier and more promising to take the second law as universally valid and impugn the gas theory than

42. Ibid., 208–18.
43. Cf. Kuhn, *Black-body theory* (1978), 20–28.

to save the gas theory by invoking "suppositions about the initial state of the universe that can hardly be formulated mathematically." The invocation would erect ignorance into an impassable barrier: "That is to turn away from any deeper insight." But, again, surrender of the possibility of any mechanical representation, as Zermelo demanded, would also block progress. Hence the compromise: Zermelo's considerations applied only to discrete masses as in the gas theory; behavior might be different in continua, and Planck expected it to be so. "I even believe and hope that in this way a strict mechanical interpretation of the second law will be found." He did not expect a quick or easy victory.[44]

By 1897 Planck had come to regard the problem of the reconciliation of mechanics and thermodynamics as the most important problem then confronting physics. He recognized his to be a minority view; in all of Berlin, he told Graetz, only he and Zermelo took a serious interest in the problem. A few years later, the official historian of the great international physics meeting in Paris in 1900 judged that not many more than four physicists in the world troubled themselves about the conflict between atomism and entropy.[45] Again Planck persevered in his unpopular quest. In 1897 he was deep into a fresh set of problems that he hoped would lead to a satisfactory representation of entropy on the basis of continuum mechanics. His hope was realized in a Pickwickian sense: within three years he had come to a resolution of the apparent opposition between mechanics and the second law and to the creation of a real one between received physics and the consequences of his own work.

The new problems concerned the equilibrium of cavity radiation. Here, in contrast to the gas theory, which oper-

44. Planck to Graetz, 23 May 1897 (DM), partially given in Kuhn, *Black-body theory* (1978), 27–28, 265–66. Cf. Planck, *Thermodynamik* (1905), v (1897).
45. Planck to Graetz, 23 May 1897 (DM); Emil Picard, "Sciences" (1903), 32.

ated with discrete material particles, the physical system under study consisted of continuous electromagnetic waves. It provided a fundamental test—so Planck explained his problem choice to a colleague—of both Maxwell's theory and the doctrine of energy.[46] Planck expected that Maxwell's equations and thermodynamics would compel any distribution of radiant energy in the cavity to a quick march to equilibrium without the backsliding allowed by Boltzmann's probabilities. We already know that the project failed and that to obtain a theory of his successful radiation formula Planck had to treat the entropy of the cavity resonators from Boltzmann's point of view. With this conversion—and the word is apt—Planck became an enthusiastic atomist and an outspoken opponent of those who still believed that the atomic theory was inimical to the progress of science.

To obtain agreement between his compromise formula for cavity radiation and his application of Boltzmann's procedures, Planck required two assumptions that in retrospect were revolutionary. First, he computed the probabilities for various distributions of energy among his resonators differently from Boltzmann's parallel calculation regarding gas molecules. Second, he divided the total energy shared among *all* the resonators at a given frequency v into units ϵ_v proportional to v, whence the famous expression $\epsilon_v = hv$. It later emerged that the first assumption introduced a statistics descriptive of a new entity, the photon or light particle, that was unlike classical molecules and resonators, and that the second had to be modified to refer to an individual resonator, not to a collective. Einstein and others recognized the necessity of this restrictive form before Planck did; in it the energy of a resonator must be either zero or an integral multiple of an "energy quantum" hv.[47] It need scarcely be

---

46. Planck to V. Bjerknes, 31 Mar. 1898 (Oslo Univ. Library).
47. Kuhn, *Black-body theory* (1978), 97–110, 170–205.

said that this quantization conflicted with the concept of energy with which Planck had worked, and which, according to his old professor, was the final capital discovery in theoretical physics. In ordinary or classical physics, energy is a continuous quantity and can be given to a resonator as ordinary people drink beer, in any quantity they wish; in Planck's theory as reinterpreted by Einstein, the resonator can possess energy only in certain amounts, as a guzzler who insists on consuming only by the pint. Why nature guzzles rather than sips became a fundamental problem for physicists.

By 1906 or 1908 Planck had come to see that his compromise over cavity radiation had loosed something brand new and menacing into the world of physics. He then sought an interpretation of the constant h that minimized the dislocation forced on physics by the discovery of the restriction on resonator energy. In 1910 he expressed himself in the manner of a protector of a menaced and even losing cause: "The introduction of the quantum of action h into the theory should be done as conservatively as possible, i.e., alterations should only be made that have shown themselves to be absolutely necessary." By then conservatism, always congenial, had become a duty for Planck under a general rule of his own devising. As scientists age and gain authority, he said, they must display "an increased caution and reticence in entering into new paths."[48]

CONSTANCY IN FLUX

The puzzling h was the definitive value of one of the two universal constants in Planck's successful radiation formula. The other, which Planck designated by k, relates mechanical measures of energy to temperature measures. He showed

48. *PA*, 2, 247 (1910), and ibid., 3, 314 (1906), respectively. Cf. *PPA*, 17 (1913).

that k also relates entropy to probability and, in indication of his debt, called it "Boltzmann's constant." It was a tribute he came to regret.[49] Although a model of modesty, Planck's sense of justice extended to himself; and the definition of k, together with the possibility of calculating it exactly from the Reichsanstalt's measurements of cavity radiation, were the first, and for several years the only, fruits of his new theory.

Whereas in 1900 h had no obvious meaning, k had a fundamental role in atomistics. The usual kinetic theory related the product LkT to the measurable pressure of a standard amount of a perfect gas (L, Loschmidt's or Avogadro's number, designates the number of molecules in the standard, and T represents the absolute temperature). The values of L in circulation in 1900 differed by a factor of ten or more. Since a knowledge of L fixed the values of many other fundamental quantities of kinetic theory, including the absolute weights of molecules, Planck's successful treatment of radiation equilibrium unexpectedly provided most important information about atoms. And it did more. From measurements on electrolysis, the total charge Q required to neutralize the electricity of L ions was known. Planck could therefore calculate the charge on a single ion, Q/L; and this charge, in accordance with the then novel theories of J. J. Thomson, should equal in magnitude the charge e carried by the cathode-ray particle or electron. The constant k, unexpectedly recovered in Planck's radiation theory, gave a clue to the connections between electrodynamics and the atomic hypothesis or, to use the terms of the times, between aether and matter.[50] It was a grand and scarcely precedented achievement.

49. *PA*, *3*, 395 (1948); Planck to E. Freundlich, 30 Mar. 1924 (MPG), re "*k* the so-called Boltzmann constant." Cf. Planck, *Theorie der Wärme* (1930), 189, and the review by F. Henning, *Nwn*, *18* (1930), 882.

50. *PA*, *1*, 728–30 (1901), 743 (1902); cf. Kuhn, *Black-body theory* (1978), 110–13; Klein, *Physics today*, *19* (1966), 23–32; and Klein, *Paul Ehrenfest* (1970), 218–24.

The calculation of the atomic constants had along with its practical value the merit of supplying an unusually strong argument in favor of the objective existence of atoms. For Planck the calculation and its underlying analysis showed that the concept of energy and the second law of thermodynamics had a deeper grounding than he had suspected. They did not stand alone and independent: their roots lay in probability theory, which in turn supposed the existence of fundamental particles. "Irreversibility leads necessarily to atomism."[51] Having undergone an irreversible change himself, Planck criticized theorists who tried to formulate their problems in a general mathematical way, without recourse to definite assumptions about the molecular constitution of matter.[52]

His confirmation of atomism and his calculation of constants were no doubt what Planck had in mind when he told his son around the turn of the century that his work on radiation would rank him among the great discoverers in physics.[53] They certainly were the accomplishments for which Planck was considered for the Nobel prize in physics in 1907 and 1908. In the latter year he almost got it: the committee for the physics prize of the Swedish Academy of Sciences put him forward, praising him particularly for having made the atomic view of nature "highly probable." The committee had debated whether to propose a division of the prize between Planck and Wien; but Arrhenius, then the dominant voice in choosing winners in physics and chemistry, had used Planck's investigations of constants and atoms to argue

51. Planck, *Thermodynamik* (1905), vii, and *Erhaltung* (1908), 64n; Planck, *Acht Vorlesungen* (1910), 40 (quote).
52. For example, *PA*, 2, 2 (1902), re the dispersion theory of Paul Drude and Woldemar Voigt; and ibid., *3*, 298–99 (1906), which places the strength of a physical theory not in the generality of its equations but in its special concepts.
53. Born, Royal Society of London, *Obituary notices*, 6 (1948), 170–71; Andrade, *Nature*, *161* (1948), 284; AH, 29.

that his contribution went well beyond an improvement of Wien's radiation formula.[54]

Arrhenius's vigorous support of Planck was but a part of a program to certify the atomic theory by a judicious bestowal of Nobel prizes. In his formal recommendation of Planck, Arrhenius stressed the agreement between the value of the electronic charge calculated from k and the value found in 1908 by Ernest Rutherford and Hans Geiger, who had counted ions (alpha particles) individually. "In this way," Arrhenius wrote, "it has been made extremely plausible that the view that matter consists of molecules and atoms is essentially correct. . . . No doubt this is the most important offspring of Planck's magnificent work."[55] That same year, 1908, the Nobel prize committee for chemistry put forward Rutherford. Unfortunately for Planck, Arrhenius's carefully concocted scheme failed to carry the academy as a whole. Although Rutherford received the chemistry prize, to the great surprise of many people—including himself—who thought him a physicist, the corresponding prize for Planck, which would have made clear the rationale for the transmutation, was not forthcoming.[56]

The majority in the Swedish Academy felt the force of one or both of the following objections: Planck produced his successful formula only by building on Wien's work and ought therefore not to have the prize alone; the formula, however well confirmed by experiment, still lacked a satisfactory the-

54. "Protokill Vid . . . Nobelkommittés för fysik sammenträde," 1908, 84 (Nobel Archives, Kungl. Vetenskapsakademien, Stockholm); Nagel, in Bernhard et al., eds., *Science* (1982), 359–61.

55. Nagel, in Bernhard et al., eds., *Science* (1982), 362. One of the few remaining objectors to atomism, Wilhelm Ostwald, formerly a confederate of Arrhenius's and van't Hoff's in establishing the ionic theory of solutions, received the Nobel prize in chemistry in 1909, shortly after stating publicly that he had come to believe in atoms.

56. Crawford, *Isis*, 75 (1984), 503–22, and Crawford, *Beginnings* (1984), 128–36.

oretical basis. The second objection invoked the authority of Hendrik Antoon Lorentz, professor of theoretical physics at the University of Leyden, whom everyone capable of judging admired as one of the greatest physicists of all time. Here is Einstein's estimate of his stature: "Everything that emanated from his supremely great mind was as clear and beautiful as a great work of art; and one had the impression that it all came out so easily and effortlessly. . . . For me personally he meant more than all the others I have met on my life's journey." Lorentz was to mean much to Planck, too. But for the moment his authority did Planck an unintentional disservice. Lorentz's proof, delivered in a lecture to mathematicians in Rome in April 1908, that ordinary physics did not lead to Planck's formula nonplussed the Swedish Academy. No one in Stockholm had known that "Planck ha[d] put forward a new, previously unimagined thought, the thought of the atomistic structure of energy."[57] The academy did not feel ready to annex energy to atomism and awarded its physics prize for 1908 to Gabriel Lippmann of Paris for inventions in color photography.

Planck's own choice for the prize for physics for 1908 was Rutherford. It was a measure of Planck's reputation for scholarship and impartiality that the Swedish Academy invited him to submit recommendations each year from the institution of the prize in 1901 through 1908. His suggestions betray none of the nationalistic bias common at the time: he nominated Wilhelm Conrad Röntgen once, in the year he won (1901); Lord Rayleigh three times, until he won (1902–4); Boltzmann twice, until he died (1905–6); Rutherford twice, until he won (1907–8). After discovering the tarnish on Planck's radiation formula, the academy did not again invite him to nominate until 1911, when he put for-

57. Einstein, in de Haas-Lorentz, *H. A. Lorentz* (1957), 8; Nagel, in Bernhard et al., eds., *Science* (1982), 363–64; Kuhn, *Black-body theory* (1978), 190–95.

ward Walther Nernst, and then not again until 1916, when he chose Johannes Stark.[58] Both eventually won. Planck's turn came in 1919, after the quantum theory deriving from his work had given rich harvests in many fields and after dozens of physicists had urged his claims on the prize givers of Stockholm.

Since Planck thought that he had solved the problem of cavity radiation in 1900 and 1901, he did not give much attention to it again until 1908, after Lorentz had forcefully brought out the discrepancy between ordinary physics and Planck's theory. That October he wrote Lorentz that he now took as the foundation of his system the restriction of resonator energy to integral multiples of hv; a few months later he suggested a rationalization of the restriction by reference to the atomism grounding irreversibility. He argued that radiation possesses entropy but not matter; hence the atomism in question must refer to energy: "In heat radiation certain energy elements play an essential role." That was not so revolutionary a thought as might appear, however, for the atomism did not have to refer to radiation processes in vacuo but only to that "still most mysterious region," the never-never land where aether and matter connect.[59] As long as it was at all plausible to do so, Planck continued to confine the difficulties he had introduced into physics to the interaction between radiation and electrons.[60]

By 1910 Einstein, Nernst, Stark, and others had applied

58. Data from the Nobel Archives, Kungl. Vetenskapsakademien, Stockholm; for nationalist bias in the nominations, see Crawford, *Beginnings* (1984), 101–8.

59. Planck to unnamed colleague, 30 Apr. 1900 (Museum Boerhaave, Leyden); Planck to Lorentz, 7 Oct. 1908 and 16 June 1909, in Kuhn, *Black-body theory* (1978), 304–5; Planck, *Acht Vorlesungen* (1910), 44–45 (quote); cf. Hermann, *Frühgeschichte* (1969), 47–56.

60. Planck to Wien, 12 July 1914 (AHQP); Sommerfeld, *Nwn*, 5 (1918), 197–98.

the concept of energy quanta far beyond cavity radiation. It appeared to Nernst, by then Planck's colleague at the University of Berlin, that the time had come for a summit meeting on the problems of radiation and quanta. It appeared to Planck, still uncomfortable with what he had wrought, that a summit would be premature: there were too few facts bearing on the problems, he told Nernst, and too few physicists conscious of the "pressing need for reform." Nernst did not discourage any more easily than Planck. He convinced the Belgian industrialist Ernest Solvay, who liked to play the philosopher, of the pressing need for a conference. Solvay underwrote the cost of a deliberative "council" of twenty-one of Europe's leading physicists, who gathered in Brussels in November 1911. The experimentalists of Berlin—Rubens, Nernst, and Emil Warburg—confirmed that Planck's radiation formula still fit all the facts and that its extension to the specific heats of solid bodies closely reproduced their measurements. Lorentz, Einstein, Sommerfeld, and other theorists stipulated that "Planck's constant" h heralded something brand new in physics, although they did not agree where to locate the novelty.[61]

Planck himself argued that the breach between quantum and ordinary physics had become too wide to be spanned by received ideas. Although, as he had expected, the Solvay conclave did not resolve the questions it was convened to discuss, it was for him an unforgettable experience. It marked the establishment of a new, expansive field of physics based upon his own pioneering work, and it showed him the personal respect in which the world's physicists held him. The meeting was intense and fatiguing, and Planck was relieved when it ended; but he would always remember the time in

61. Planck to Nernst, 11 June 1910, in Hermann, *Frühgeschichte* (1969), 153–54. Cf. Kuhn, *Black-body theory* (1978), 230.

Brussels as scientifically exciting and personally most gratifying.[62]

Enthusiastic Relativist

For several years before resuming his study of the quantum theory, Planck had been occupied in extending the theory of relativity published by Einstein in 1905. Einstein may be considered Planck's second great discovery in physics; Planck was the first important theorist to follow Einstein's lead, and his support, in Einstein's judgment, was instrumental in securing the swift acceptance of the new ideas among physicists.[63] In the first colloquium of the winter semester 1905–6, Planck reviewed the theory of relativity and corrected a slip in Einstein's original presentation;[64] and in 1906 he defended it before the Naturforscherversammlung against the apparently disconfirming experiments of Walther Kaufmann.

Kaufmann's technique and measurements, Planck said, could not decide between the various theories then current that predicted changes of mass with velocity.[65] All these theories except Einstein's postulated an electron model: spheres rigid or deformable, with charge distributed throughout or on the surface, all arbitrary pictures and all distasteful to Planck. Einstein's theory had no need of such hypotheses; moreover, it had the special merit of doing without a special

62. Planck, in Institut International de Physique Solvay, *Théorie du rayonnement*, 108–9; Planck to Wien, 8 Dec. 1911 (Autogr. I/285, SPK), and to Lorentz, 11 Oct. 1913 (LTZ/4).

63. Einstein, *Nwn, 1* (1913), 1079; Born, *My life* (1978), 130, recalled being inspired to study relativity by news of Planck's endorsement of it.

64. Laue, *Nwn, 35* (1948), 3.

65. *PA, 2,* 121–35 (1906), 210–14 (1907); cf. Goldberg, *HSPS, 7* (1976), 127, 131–32, 154.

coordinate system. At the time Planck thought that relativity deserved to be the first item on the agenda of theoretical physics merely for the simplification it introduced into the electrodynamics of moving bodies. Later he spoke of the "relief" the establishment of Einstein's ideas brought to perplexed electrodynamicists.[66]

The paradoxes of relativity—the contraction of space, the dilation of time, the mixing of space and time—apparently did not bother Planck. Indeed, they appealed to him: as he later explained, the paradoxes arise only because of human sensibilities, and the physicist's ability to transcend even such deep-rooted intuitions as those of space and time supports the hope that mankind can achieve a truly universal physics, as acceptable to Martians as to human beings. About the same time Planck urged a similar nostrum to aid in the digestion of the probabilistic interpretation of the second law: by freeing the theorist of the old-style thermodynamic analysis, in which thought experiments played a major part, Boltzmann's method deanthropomorphized the concept of entropy, emancipated it from "human experimental technique," and raised the second law to a "real principle."[67]

One of the striking oddities of relativity theory is that the usual rule for the addition of velocities along a straight line, $w = u + v$, no longer holds. Its violation, which followed from Einstein's postulate that the speed of light in free space is the same to all observers, indicated a need to revise the sacred mechanics of Newton, of which the usual addition rule was a direct consequence. Planck undertook to perform this sacrilege. As early as 1906 he could show in a simple case how to rewrite Newton's laws of motion in the spirit of relativity.

66. *PA*, *2*, 115–16 (1906); Planck, *New science* (1959), 50.
67. Planck, *Acht Vorlesungen* (1910), 43; cf. *PA*, *1*, 380–81 (1891).

In the next two years Planck worked out a relativistic dynamics based upon the principle of least action, a general formulation of ordinary mechanics that has the singular advantage that it can be expressed in the same form in all relativistically equivalent reference systems. The laws of energy and momentum conservation do not enjoy this relativistic invariance. "Over both of them reigns the principle of least action, which seems to control all the reversible processes of physics."[68] Least action had fascinated philosophical physicists by its hint of purpose or design since its first formulation in the middle of the eighteenth century. (It requires that every mechanical process take place so that the "action"—a mathematical function of the momenta and positions of the participating particles—has a minimum value. The teleologist may wish to regard it as a precise formulation of the dictum, *natura nihil facit frustra*.) Planck's teacher and colleague Helmholtz had shown the wide applicability of the principle of least action, and Planck in his turn was following the path it opened "to a unified conception of all the forces of nature."[69] No doubt the teleological undertone in the principle's formulation as well as its historical pedigree also recommended it to him.

In developing at large the consequences of relativity and least action, Planck began with the one physical system whose several properties—dynamic, electrodynamic, and mechanical—could be stated with absolute precision and also independently, he said, of special hypotheses. It was his very own system of radiation locked in a cavity, which he apparently still regarded in mid-1907 as obedient to ordinary physics. His development brought, among other things, a general proof of the famous relativistic equivalence of mass and energy, $m = E/c^2$; an argument that entropy, like the probabilistic computation that defines it, must be a relativ-

68. *PA*, *3*, 100–101 (1914).
69. Ibid., 323 (1906). Cf. Planck to Einstein, 6 July 1907 (AE).

istic invariant; transformation equations for pressure and temperature; and a list of invariants. In this connection Planck pointed to a most gratifying consequence of the invariance of the physico-historico-teleological principle of least action. "If the law that a completely definite elementary quantum of action exists be taken into account . . . , one can also say: a definite number of action elements, which is independent of the choice of reference system, corresponds to every change in nature."[70] Evidently the number of quanta must be invariant, and their measure, Planck's quantum h, must have a significance truly profound.

The personal connection between Planck and Einstein was to be as complex and problematic as the relations between quanta and relativity. Planck had the very greatest admiration for Einstein's work, which he liked to compare with Copernicus's.[71] Einstein at first judged Planck to be more like Copernicus's opponents, "stubbornly attached to preconceived opinions that are undoubtedly false." This was in 1911, after the two had met at the Solvay council. In 1913 Einstein disparaged Planck and all the other leading German theorists for resisting his generalization of relativity: "Laue [Planck's favorite student] cannot be reached by fundamental considerations, nor Planck. . . . A free, unconstrained view is not the mark of an (adult) German (blinders!)."[72] But, as Einstein allowed, resistance itself might be a strength if, as in Planck's case, it arose from adherence to very general principles and their confirmed consequences. "This is char-

70. *PA*, 2, 180–83, 188, 191–92, 198 (quote), 202–7 (1907); Goldberg, *HSPS*, 7 (1976), 137–41. Cf. Planck to Einstein, 6 July 1907, in Hermann, *PB*, 25 (1969), 434.

71. An early example is Planck to Felix Ehrenhaft, 1908, in "Max Planck zum Gedächtnis," *PB*, 4 (1948), 169–70.

72. Einstein to Heinrich Zangger, 16 Nov. 1911, in Seelig, *Helle Zeit* (1956), 43; to Michele Besso, late 1913, in Einstein and Besso, *Correspondance* (1972), 50; Planck to Wien, 12 July 1914 (AHQP). Unknown to Einstein, Planck supported a proposal to test general relativity as early as 1912; Planck to Karl Schwarzschild, 21 Jan. 1913 (AHQP).

acteristic of his entire way of working, perhaps of the method of the pure theorist in general."[73]

In 1913 Planck and Nernst put together a set of enticing sinecures that would bring the mercurial and unconventional Einstein into the staid circle of Berlin physicists: an ordinary membership at high salary in the Berlin Academy, a professorship without teaching obligations at the University, the directorship of the nonexistent Kaiser-Wilhelm Institute for Physics.[74] During the First World War Planck and Einstein came to esteem each other personally and to draw together as physicists. After careful study and self-persuasion Planck accepted the general theory of relativity, "however only as a theorist." "He is a splendid fellow," Einstein wrote his old friend Michele Besso. "My other experiences with colleagues show a terrible preponderance of the all-too-human!"[75] Their relationship proved strong enough to withstand the divisive force of their sharply antagonistic political positions. But ultimately the all-too-human could not be denied, and their warm personal and collegial relations during the Weimar regime did not prevent a final break under the Third Reich.

Concerned Pedagogue

Whereas Einstein soon separated from his first wife and his two children left to live with their mother, Planck raised

73. Einstein, *Nwn*, *1* (1913), 1077.

74. Kirsten and Körber, *1*, 201–3; Kirsten and Treder, *1*, 95–105. Cf. Planck to Wien, 31 July 1913 (AHQP).

75. Respectively, Planck to Wien, 30 Oct. 1919 (AHQP), and Einstein to Besso, 21 Dec. [1915], in Einstein and Besso, *Correspondance* (1972), 61. Cf. Einstein, in Hermann, *New physics* (1979), 34, and Oseen to Bohr, 3 Mar. 1916, in Bohr, *Works* (1972), *2*, 570: "What constitutes Planck's great strength seems to me to be that he aims at such a logical purity."

a family of four in perfect contentment until his first wife died in 1909, after twenty-three years of marriage. Planck, then fifty, remarried a year and a half later; although sadly afflicted by his wife's death, he needed another wife for his house and children, for companionship, and because a professor customarily had one. His second wife, Marga von Hoesslin, a niece of his first, was a source of great strength to him. Einstein, who was thirty-five when he separated from his first wife, lived alone until 1919, when he married his cousin, mainly for the convenience of having his shirts done at home.[76] Planck, rightist and imperialist, was a family man; Einstein, leftist and humanitarian, was a loner. Einstein liked rough joking, which he did not disdain to indulge in with students and journalists. Planck, although sociable, was very reserved and could enjoy himself only with people of his own rank, with whom he might take a glass of wine and a cigar and even a quiet joke with great pleasure.[77] But he always preferred the company of his family. "How wonderful it is to set everything else aside," he had written as a young man, "and live entirely within the family." There, and not with his fellows, he allowed his warm nature full play. Marga Planck noticed this characteristic: "He only showed himself fully in all his human qualities in the family."[78]

Planck settled his family in a large house with a spacious garden built to his specifications in countrified Grunewald, a Berlin suburb favored by university professors. His close neighbors and colleagues included Hans Delbrück, an emi-

76. Planck to Runge, 28 July 1909 (RP), and to Emma Lenz, 27 Oct. 1909 and 19 Oct. 1910 (MPG); AH, 20, 39, 45; W. H. Westphal, in "Max Planck zum Gedächtnis," *PB*, *4* (1948), 168; Clark, *Einstein* (1971), 176–77, 193, 220–21.

77. Marga Planck to Ehrenfest, 26 Apr. 1933 (ESC); Gerlach, *Die Quantentheorie* (1948), 29.

78. Planck to Runge, 31 July 1877 (RP); Marga Planck to Einstein, 1 Feb. 1948 (AE).

nent military historian, who was to help moderate Planck's political views during the First World War; and Delbrück's brother-in-law, the theologian-historian-administrator Adolf von Harnack, with whom Planck would work closely on the organization of German science.[79]

Planck's villa was full of music. He had the piano technique of a professional musician. As a student he had composed songs and an entire operetta for musical evenings in professorial houses; he served as second choir master in an academic singing group, played the organ at services in the students' church, conducted an orchestra, and studied harmony and counterpoint. At performances at his home he might accompany his good friend, the great violinist Joseph Joachim, or play in a trio that included Einstein; or he might lead friends, neighborhood children, and his twin daughters—who had inherited his musicality—in choral singing. Before the war this motley singing society met every two weeks. In those days Planck's sense of pitch was so perfect that he could scarcely enjoy a concert, much less the neighborhood children; but, like his politics and his thermodynamics, his ear gradually lost its absolutism and allowed him greater satisfaction.[80]

Planck's own playing was not only a means of relaxation and recreation but also, as his nephew-in-law Hans Hartmann wrote, the only domain of life in which he gave his spirit free rein. Although he played the music of all great composers, he preferred Schubert and Brahms to Bach, and he admired Schumann; of Bach he singled out the pathetic and emotional parts of the Saint Matthew Passion: all

79. Westphal, *Nwn, 45* (1958), 234, and in "Max Planck zum Gedächtnis," *PB, 4,* (1948), 167; Agnes von Zahn-Harnack, *Harnack* (1951), 343.

80. Planck to Max Schirmer, 26 Dec. 1923 (MPG); AH, 23–25; Hermann, *New physics* (1979), 34; HH, 11; reminiscences by Laue, *Nwn, 45* (1958), 223; von Zahn-Harnack, in "Max Planck zum Gedächtnis," *PB, 4* (1948), 165–67; Hahn, MPG, *Mitteilungen* (1957), 243.

choices that reveal the deep romantic vein beneath his still exterior.[81] This romanticism was of a piece with his quest for a super- or transhuman world picture and with his sense of tradition and country.

Planck's world centered on the chief institutions of learning in the chief center of learning in Germany: the University and the Academy of Berlin. To these he added the leading specialist society for physics in the world, the German (formally the Berlin) Physical Society, which controlled the world's leading physics journal, the *Annalen der Physik*. The University had a colloquium and the Academy a meeting every week; the Society gathered once a fortnight; and Planck attended them all. In addition he had the usual duties of a professor at a large university, "the eternal meetings, examinations, and writing of reports."[82] For this he received a salary of 4,400 marks and a housing supplement of 900. This sum, together with his salary of 900 marks from the Academy, gave him a base income (exclusive of fees and royalties) from his profession of 6,200 marks, slightly below the average for ordinary professors of physics around 1900.[83] No doubt the fees and royalties, which brought in a useful amount, and also, perhaps, income from his wife were necessary to run the pleasant villa in Grunewald.[84]

Planck's sense of duty and deliberateness of action made

81. HH, 11–12; Planck to Schäfer, 24 Dec. 1935 (Nachlass Wilhelm Schäfer, Heinrich-Heine-Institut, Düsseldorf), and to Laue, 15 Feb. 1942 (DM). Cf. Westphal, *Nwn*, *45* (1958), 234.

82. Planck to Sommerfeld, 11 Sept. 1899 (AHQP 33/3), excerpted in Kangro, *Vorgeschichte* (1970), 148; Planck to van't Hoff, 25 Dec. 1894, and to F. A. F. C. Went, 4 Feb. 1928 (Museum Boerhaave, Leyden); and to Emil Fischer, 14 Dec. 1907 (acc. Darmst. 1917.141, SPK).

83. Planck to Fischer, 8 May 1895 (TBL); Forman, Heilbron, and Weart, *HSPS*, *5* (1975), 40–42.

84. Royalties from Planck's many textbooks and essays averaged perhaps one thousand prewar marks a year or more; one essay of 1919 brought almost seven hundred marks in its first edition. Davidis, ed., *Wissenschaft* (1985), 54–55.

him a leading citizen, even a guardian, of the University. He took on the powerful Prussian education minister, Friedrich Althoff, over the choice of the successor to August Kundt, professor of experimental physics, who died in 1894. The succession was most important: not only did it concern the center of Prussian learning, the University of Berlin, but also a subject that had recently enjoyed a rapid increase in enrollment owing to its growing relevance to technology.[85] The faculty recommended Emil Warburg; Althoff went back and forth between other candidates, "only [in Planck's opinion] to evade the faculty's recommendation." That irritated Planck. "One need not be pro-Semitic [Warburg was Jewish] to find such a procedure highly questionable, which fundamentally disregards the objective considerations . . . and especially the authority of the faculty, which reached its conclusions after careful deliberation." Planck looked for an opportunity to open his mind to the minister. Perhaps he found it: Warburg, who shared Planck's general view of the world, was appointed.[86]

Another example of Planck's academic stewardship was his stand in the famous Arons case, which pitted privileges of the University against duty to the Prussian government. As a member of a special commission appointed by the University in 1895, Planck helped to respond to the order of the Prussian Kultusministerium (Ministry of Public Worship and Education) to discipline a *Privatdozent* (lecturer) in physics, Leo Arons, whose offense was to be a member of the Social Democratic party and to support it with speeches and money. Arons had a good supply of both, being an academic and—like Planck—the son-in-law of a banker. For a long time he subsidized the party's chief publication, the *Sozialdemokratische Monatshefte*.

85. Planck to Kultusminister, 14 Mar. 1895 (acc. Darmst. 1913.51, SPK).

86. Planck to Fischer, 21 Oct. and 18 Nov. 1894 (TBL).

According to law and custom, the right to discipline *Privatdozenten*, who were not civil servants, resided in the University, not the ministry. Planck had earlier praised Arons for his dedication to keeping the laboratory exercises going while the ministry sought Kundt's successor;[87] Planck's commission found that Arons was a good teacher and competent scientist and that he did not inject his political views into the classroom. The commission declined to discipline him. Its decision prompted a sharp exchange of journalistic opinion, conservative newspapers condemning the faculty for its "reckless behavior," liberal ones praising it as "the defender of the freedom of science." In 1898 the Prussian legislature passed a law, known familiarly as the lex Arons, that assigned disciplinary authority over *Privatdozenten* to the Kultusministerium. The minister again asked the University to expel Arons; Planck's commission stood fast, and the government had to do its own dirty work.[88] It could not have been pleasant for Planck to oppose his government, but when his precious republic of letters came into conflict with it, he knew where to stand. At least he did so around 1900.

In other contemporary university business Planck also showed himself on the liberal side. Responding to an enquiry put to many prominent academics in 1897, Planck allowed that women should not be denied in principle the right to study at a university. He had tried the experiment himself, "as far as it is compatible with academic order," and had given a few women revocable permission to attend his lectures. He thus put into practice the sympathy for women's

87. Planck to Kultusminister, 14 Mar. 1895 (acc. Darmst. 1913.51, SPK).

88. Fricke, *Zeitschrift für Geschichtswissenschaft*, 8 (1960), 1083, 1087–88, 1101–2. Cf. Ringer, *Decline* (1969), 55, 141–42, and the memorial issue for Arons in the *Sozialistische Monatshefte*, no. 25–26 (1919), 1059–74, which includes an appreciation by Einstein of Arons as a physicist.

rights awakened in him years earlier by the heroine, Nora, in Ibsen's *A doll's house*, whose captivity in a male world offended his sense of justice. Still it would not do, he said, to entice women from the plane fixed for them by their social and biological duties. "In general it cannot be emphasized strongly enough that Nature itself has prescribed for a woman her place as mother and housewife and that natural laws cannot be ignored under any circumstances without serious damage, which in this case would appear especially in the next generation." The courage to defy nature in this particular gradually waxed among German academics. In 1913–14, in Planck's year as rector of the University, no fewer than 770 women were studying in Berlin, and throughout the Reich they numbered almost 3,500, some 6 percent of the university population.[89]

By then Planck had become a champion of the exceptional woman scientist. He had met one in that chamber of surprises, the institute for theoretical physics of the University of Vienna, where the statistical rendering of the second law developed. Planck was asked to succeed Boltzmann, who had died a suicide in 1906. Among the attractions of the place—endless music, proximity to Planck's favorite mountain hikes, and the promise of a new institute—was Boltzmann's student Lise Meitner. Planck inclined toward taking the place of the man whose work he had learned to admire as "one of the most beautiful triumphs of theoretical research," but he at last declined, primarily, he told Wien, because of the unexpected interest the Berlin faculty showed in retaining him.[90] Meitner then went to Berlin, where with

89. A. Kirchhoff, ed., *Die akademische Frau* (1897), 256–57, reproduced in Kleinert, *PB*, *34* (1978), 32; Runge, *Carl Runge* (1949), 34, on Nora; *Berliner akademische Nachrichten*, *8* (1913–14), 62; Lenz, *Geschichte* (1910), *3*, 498.

90. *PA*, *3*, 227 (1932); Planck to Wien, 19 June 1909 (AHQP); Meitner, *Bulletin of the atomic scientists*, *20:11* (1964), 3; Westphal, in "Max Planck zum Gedächtnis," *PB*, *4* (1948), 168.

Planck's help she obtained a place in the chemical institute of his fellow professor and academician Emil Fischer. In 1912 Planck appointed her his own assistant, and two years later, together with Fischer, he defeated an effort to draw her away to the University of Prague. Meitner regularly participated in Planck's musical evenings along with her co-worker Otto Hahn. She became one of the Plancks' closest friends.[91]

Planck's pedagogical obligations increased rapidly after 1900. He advised countless contributors to the *Annalen der Physik*, with which he had become associated as an editorial overseer for the Physical Society in 1895, in succession to Helmholtz. Planck's involvement with the journal probably increased after 1906, when Wien took over the day-to-day editing in succession to Paul Drude. (Drude had succeeded Emil Warburg, who went to head the Reichsanstalt in 1905; Drude's suicide in 1906 brought Rubens to Kundt's old professorship, which stabilized the Berlin experimentalists for a decade and a half.)[92] Planck advised on the disposition of manuscripts on theoretical topics; in the process he developed clear demarcations between physics on the one hand and mathematics and philosophy, which he considered its auxiliaries, on the other. These demarcations and Planck's opposition to mathematical showpieces helped to define the scope and method of the emerging specialty of theoretical physics.[93]

From 1894 to 1905 Planck had the general supervision of

91. Hermann, *New physics* (1979), 41; W. R. Shea, "Introduction," in Shea, ed., *Otto Hahn* (1983), 11; Planck to Fischer, 17 May 1914 (TBL); Meitner, *Nwn*, *45* (1958), 406; Planck to Wien, 25 Aug. 1925 (AHQP): "I value her very highly not only scientifically but also personally, she is a true friend."

92. Cf. D. Hoffmann, AW, ITGOW, *Kolloquien*, *35* (1984), 5–30.

93. Pyenson, *Young Einstein* (1985), 164–65, 197, 200, 203–5, 208; Planck to Wien, 15 Oct. 1906, 22 Jan. 1923, and 13 Oct. 1924 (AHQP). Planck to Leo Koenigsberg, 2 June 1899 (acc. Darmst. 1922.93, SPK), mentions the difficulty of finding theorists with just the right amount of mathematics.

the University's institute of experimental physics during Warburg's tenure of the professorship (to reassure the ministry?) as well as of his own small institute for theoretical physics, to which an assistant and a library were attached. The greatest demands arose from the sharp increase in enrollments in his lectures: 18 in 1890, 89 in 1900, 143 in 1909. He lectured four times a week and on the fifth day presided over problem solving. He went through the entire course of lectures—mechanics, hydrodynamics, electrodynamics, optics, thermodynamics, kinetic theory—once every six semesters.[94] Eventually he published books covering most of the topics treated, and his texts on thermodynamics and heat radiation, which are models of clarity, had several editions in several languages.

The treatment was of course orderly and unified. People used to the disjointed and often bumbling English style of instruction were particularly impressed. The chemist J. R. Partington judged Planck to be the best lecturer he had ever heard. The Indian physicist D. M. Bose, who had studied in Cambridge, experienced what he called a "revelation" in Berlin. "After attending Planck's lectures, [I knew] what a system of Physics meant in which the whole subject was developed from a unitary standpoint and with the minimum of assumptions."[95] Perfect lectures may not, however, inspire as well as less studied ones, as Planck himself had remarked in the case of Kirchhoff's.[96] Perhaps for that reason and also because of his policy of leaving doctoral candidates on their own, as his professors had left him, Planck drew few advanced students and established no school.

Some two dozen people took doctorates under Planck in

94. Lenz, *Geschichte* (1910), *3*, after p. 446; Planck, in ibid., 276–78; Laue, *Nwn*, *35* (1948), 4.

95. AH, 47–48; Partington, in "Max Planck zum Gedächtnis," *PB*, *4* (1948), 172; Bose, *Science and culture*, *13* (1947), 238.

96. *PA*, *3*, 376 (1948).

his forty years of teaching, during which he endured at least 650 doctoral examinations. The most distinguished were Max Abraham (1897), Max von Laue (1903), Moritz Schlick (1904), Kurt von Mosengeil (1906), Walther Meissner (1906), Friedrich Reiche (1907), Ernst Lamla (1912), Walter Schottky (1912), Walther Bothe (1914), and perhaps one or two more. This list contains two Nobel prize winners (Laue and Bothe), several other important physicists (Abraham, Meissner, Reiche, Schottky), and a well-known philosopher (Schlick). Mosengeil, one of the most promising, died in a climbing accident soon after finishing his thesis, which Planck saw through the press.[97] The thesis concerned relativity, on which at least a third of Planck's doctorands worked, mainly between 1906 and 1914. Some students began their climb up the academic ladder by qualifying as *Privatdozenten* under Planck: Otto Kriegar-Menzel (1894), writing on mechanics; Albert Byk and Laue (1906), on thermodynamics; Reiche (1913), on radiation theory; Rudolf Seeliger (1915), on the theory of electric discharges; Gerhard Hettner (1921), on quantum theory; and Bothe (1925), on the photoelectric effect. Of all these, Laue, who climbed all the lower steps of the academic ladder under Planck and made most of his career in Berlin, was his teacher's most faithful disciple and "a close and true friend."[98]

Planck's concern with pedagogy is better illustrated by his care in preparing courses than by his benign neglect of doc-

97. List in Meissner, *Science*, *113* (1951), 80, correcting Born, Royal Society of London, *Obituary notices*, *6* (1948), 196–97; Pyenson, *Young Einstein* (1985), 212, lists twelve theses on which Planck was an adviser; in a note in *Naturwissenschaft-Technik-Medizin*, *21:2* (1984), 93, Dieter Hoffmann gives twenty-three as the number of Planck's doctorands, and in AW, ITGOW, *Kolloquien*, *35* (1984), 69–70, he lists twenty-five. Cf. AH, 48–49; Meitner, *Nwn*, *45* (1958), 407–8; Goldberg, *HSPS*, *7* (1976), 133–38.

98. Quote from *PA*, *3*, 415 (1938); data from D. Hoffmann, AW, ITGOW, *Kolloquien*, *35* (1984), 71.

toral students. He took a strong interest in the instruction of beginners, who might easily lose the way without guidance. Even in his gymnasium days, when he had substituted for his mathematics teacher for six weeks, he had experienced the "tremendous satisfaction of teaching another person and seeing how the information falls on fruitful soil and multiplies there."[99] The chief work of his assistant was to correct beginners' weekly problem sets and discuss the results with Planck, who would ask students who had done well to demonstrate for the others during the exercise hour.[100] The technique simultaneously encouraged competition and discouraged handing in someone else's correct answers.

Planck also took an interest in reforms of physics teaching in the secondary schools and in the preparation of physics teachers in the universities. These reforms had been under lively discussion by professional groups like the Physical Society and the Naturforscherversammlung since the beginning of the century. All agreed that experimental physics should have a place in the secondary schools that corresponded to its importance for society and to "its significance for the development of the mind."[101] Most of the Prussian schools surveyed in 1905 declared themselves willing to offer practical instruction if means were provided to purchase the necessary apparatus.[102] From the university side two questions stood out: how should prospective teachers prepare for the new situation expected in the schools, and how should

99. Planck to Runge, 9 Dec. 1878 (RP).

100. Lamla, in "Max Planck zum Gedächtnis," *PB*, *4* (1948), 173; Planck, in Lenz, *Geschichte* (1910), *3*, 276–78.

101. Schreber, *ZPCU*, *19* (1906), 213 (quote); GDNA, Unterrichtskommission, *Gesamtbericht* (1908), iii–vii. Cf. Wiedemann, *ZPCU*, *19* (1906), 267, and Pyenson, *Neohumanism* (1983), 88–93.

102. Poske, *ZPCU*, *20* (1907), 401; GDNA, Unterrichtskommission, *Gesamtbericht* (1908), 243.

introductory courses at the university be taught in order to continue the education of the better-prepared students from the reformed schools?

The reformers agreed that prospective teachers should take a special course in which they learned to set up and repair apparatus and use the tools of the instrument maker; they should also practice lecturing and, perhaps, learn something of the "historical-critical" side of their subject. This last suggestion was advanced by the most knowledgeable of the group, Karl T. Fischer of Göttingen, who had been asked to prepare formal proposals by the education committee of the Naturforscherversammlung; and it had been put into practice by Heinrich Konen in Münster, and perhaps by others.[103]

As for the level of the introductory university course, many thought it aimed too low, repeating for the benefit of prospective physicians and pharmacists what the better graduates of the better secondary schools already knew. The repetition tended to drive the prepared students to mathematics courses, from which they emerged unfit to teach experimental physics in the secondary schools. Reformers hoped to break the circle by raising the level of university instruction.[104] With this plan Planck perfectly agreed, although with his usual restraint. He was present at the Physical Society when it passed a resolution that experimental lectures at the university should be directed at the level of

103. K. T. Fischer, *ZPCU*, *20* (1907), 65–66, 74–76, endorsed by Noack, ibid., 147–53; Konen, ibid., 231–34; and GDNA, Unterrichtskommission, *Gesamtbericht* (1908), 249, 268–69, 287, 296, 299. Cf. Schreber, *ZPCU*, *19* (1906), 213–15; Wiedemann, ibid., 270; Börnstein, ibid., 355–56; and Grimsehl, *ZPCU*, *20* (1907), 6.

104. Schreber, *ZPCU*, *19* (1906), 213; Wiedemann, ibid., 267; Grimsehl, *ZPCU*, *20* (1907), 3–5; K. T. Fischer, ibid., 68; GDNA, Unterrichtskommission, *Gesamtbericht* (1908), 275–76.

the gymnasia or equivalent schools.[105] As he told Wien, the level in question should be that of the average attainment of graduates, not that of the most distinguished.

It is plausible that these pedagogic questions helped motivate Planck's first public declaration of his philosophy of science, which he made in December 1908. The declaration came as an apparently gratuitous attack on the leading German-speaking epistemologist, the Austrian philosopher-physicist Ernst Mach, who continued to spurn atoms after others had given in and who insinuated that a physics free from presuppositions could be built from direct sense experience alone. These propositions challenged the position that Planck had reached during his conversion around 1900; consequently, he could not favor Fischer's proposal to make "historical-critical"—that is, Machist—philosophy a standard ingredient of the seminar for physics teachers. Planck knew the appeal of Mach's stiff independence, in which he had sensed a kindred spirit when he was fighting for room for his brand of thermodynamics. He knew the allure and, as he thought, the danger of Mach's promise of a science free from metaphysics, which had once beguiled him and in whose name his former students Abraham and Lamla resisted the developments of modern physics.[106]

Traces of Machist ideas and phraseology turn up here and there in Planck's earliest pedagogical considerations. Writing Runge in 1881, as he worried how best to present mechanics to beginning students, he insisted on the need to stay close to experience. Hence the notion of force should be introduced by reference to muscular effort, and the goal of science should be represented as the calculation of the sense

105. Planck to Wien, 9 Oct., 26 and 30 Nov. 1908 (AHQP).
106. Planck, *Physikalische Zeitschrift, 11* (1910), 1187; Goldberg, *Archive for history of exact science,* 7 (1971), 23, citing Max Abraham, *Scientia, 15* (1914), 12, 16; Meitner, *Bulletin of the atomic scientists, 20:11* (1964), 3, on Lamla. Cf. *PhW,* 151–52.

impressions that a person would experience under stated conditions.[107] In his Göttingen prize essay of 1887, he declared that the deeper, more comprehensive, and more certain a law of physics was, the closer its connection with the immediate results of observation must be. Hence the need to base the energy principle "on true facts of experience . . . , with the utmost avoidance of all hypotheses." These words come from the preface to Planck's *Prinzip der Erhaltung der Energie*, published in 1887; a decade later he still claimed the laws of thermodynamics as "very general facts."[108] But never, as we know, did Planck share Mach's opinion that mechanical reduction could not be achieved; at the very least it provided a model for the unified approach to physics that Planck held to be the greatest desideratum in science.[109]

Even before Planck had converted to Boltzmann's approach, he had unmasked Mach as a false prophet, ignorant of thermodynamics. "As far as Mach is concerned, I must say that although I otherwise much appreciate the independence and sharpness of his judgment, I do not think him competent as far as the second law is concerned."[110] In 1896 Planck felt himself obliged to speak out against Wilhelm Ostwald's energetics, which drew some of its negative arguments from Mach. Energeticists wished to replace all physics, including elementary mechanics, by the equations of energy and its transformations. Planck joined Boltzmann in condemning the program as nonsense. Typically, he motivated his sharp attack as a matter of pedagogy and duty: "I consider it my duty to warn most emphatically against the

107. Planck to Runge, 15 Sept. and 24 Oct. 1881 (RP).

108. Planck, *Erhaltung* (1908), vii–viii, xiii; *Thermodynamik* (1905), v. Cf. *PhW*, 89, 94–95.

109. Planck, *Erhaltung* (1908), 58; *PhW*, 99–100.

110. Planck to Ostwald, 16 July 1891, in Körber, ed., *Briefwechsel Ostwalds* (1961), *1*, 35.

further development of energetics in the direction it has recently taken, which signifies a serious backward step from the current results of theoretical work and can only have the consequence of encouraging young scientists in dilettantish speculation instead of in a thorough grounding in the study of established masterpieces."[111] After his investigation of radiation had converted him to Boltzmann's approach, Planck not unnaturally supposed that the length and tediousness of the effort owed something to Mach's influence, which he began to see as mischievous and even dangerous.[112]

In 1908 Planck had several occasions to reconsider Mach's philosophy. He found it explicitly in proposals to improve science teaching and in the Machist passages in the proofs of the reprint of his own *Prinzip der Erhaltung der Energie*; he met it, or rather its insufficiency, implicitly in his work on relativity theory, in which the major advance rested on rejecting the testimony of the senses. As a concerned pedagogue, as a professor of physics, he felt an obligation to speak out; as a shrewd observer of academic politics, he knew where and at whom to shout. During the fall, upset by the recognition of the trouble his quantum theory had brought to physics, by Marie's illness, and by the leak of the false rumor that he had been awarded the Nobel prize, he was in an uncharacteristically combative mood as he pondered what to say in his first major address outside Germany, scheduled for Lorentz's Leyden in December.[113]

111. *PA*, *1*, 464–65 (1896). Cf. the exchange with Ostwald in Körber, ed., *Briefwechsel Ostwalds* (1961), *1*, 42–59.

112. His strongest statement is in *Nwn*, *28* (1940), 778–79; cf. Blackmore, *Ernst Mach* (1972), 217–22.

113. Kuhn, *Black-body theory* (1978), 197–99; Planck to Wien, 30 Nov. 1908 (AHQP): "The [Nobel business] has so far brought me only irritation because of completely nonsensical newspaper reports. I ask therefore only for your silent sympathy."

| 2 |

DEFENDING THE
WORLD PICTURE

Against Mach

Planck's Leyden lecture developed around the question whether physics was approaching unity. Once answered correctly, that is, affirmatively, the question forced the philosophical physicist to confront a wider problem: is the world picture implied and realized by unification merely a representation of the facts convenient for us, or does it rather reflect objective natural processes entirely independent of the perceiving mind?[1] As for the original question, nothing could be clearer. At first physics had at least as many main branches as physicists had senses: mechanics, optics, acoustics, and the sciences of heat, electricity, and magnetism. Mechanics soon absorbed acoustics, and, latterly, heat; electricity and magnetism united, and subsumed optics. It was only a question which of the two survivors would swallow the other. Planck bet on electrodynamics. Indeed, he said, the opposition between the two had been much exaggerated. Adapting a theme he had been stressing for years, he located the only profound cleavage in physics in the apparently un-

1. *PA*, *3*, 7 (1909).

47

bridgeable chasm between reversible and irreversible processes.[2]

We know Planck believed that probabilities and helter-skelter molecules could fill the chasm. He extracted from Boltzmann's deus ex machina not only the blessing of unification but also the sacrament of an objective world picture: the probabilistic interpretation of entropy simultaneously saved the program of mechanical reduction and raised the concept of irreversibility to a principle independent of human striving and ingenuity. With that emancipation thermodynamics shared preeminently in the fundamental characteristic of the unifying process, the progressive elimination of the parochial from the world picture.[3]

Music, however beautiful it may sound, is nothing but a beating of the air; heat, but another form of motion; light, an alternation of electricity and magnetism. The goal of physics is not only conceptual unity but a unification on principles that all physicists everywhere, with or without human senses, can come to independently. As persuasive indicators that unification by deanthropomorphism can succeed, Planck cited the constants h and k that figure in his radiation law. They and the constants of gravitation and the speed of light have a universal character independent of reference to any particular body or substance. And that is the point: "With their help we have the possibility of establishing units of length, time, mass, and temperature, which necessarily retain their significance for all cultures, even unearthly and nonhuman ones."[4]

The goal of a unified, transhuman science implies that the question "Are there real natural processes entirely independent of ourselves?" must be answered affirmatively. But that

2. Ibid., 7–10, 15–16; cf. Planck to Ostwald, 24 July 1898, in Körber, ed., *Briefwechsel Ostwalds* (1961), *1*, 62.
3. *PA*, *3*, 19 (1909).
4. Ibid., 21, 23.

answer conflicts with the basic tenet of Mach's philosophy, which, as Planck represented it, placed the only reality in sensations and conceived the goal of physics to be the most economical adjustment of human thought to human experience. It appeared that the Machists chose as the basis of science precisely those elements—the particular and the parochial—that science discards as it progresses. Their doctrine made unintelligible the fact and the standard on which physics rests: independent investigators, working in different times and places, must agree about the phenomena before the experiences of any of them have any scientific value.

These were airy problems, however, and one could do physics without solving them. Much more sinister for Planck was Mach's misleading of practicing physicists by his attack on atomism. With his convert's zeal, Planck held atoms to be no less real than planets. Mach's insistence on the priority of sensations played a useful part when zealous scientists, inspired by the principle of energy conservation, taught that the entire natural world would soon be explained on the basis of mechanics. "But [Mach] overshoots the mark by degrading the physical world picture along with the mechanical."[5]

Planck concluded with an argument that Mach probably found the most offensive of the entire polemic. Mach's best-known work, *The science of mechanics, a critical and historical account*, grounds and exemplifies his epistemology with historical examples. Planck claimed that none of the great masters of physics had created their science with Mach's precepts in mind: "Economical considerations were certainly the very last thing that steeled these men in their fight against received conceptions and high-and-mighty authorities." In short, nothing ever came from physicists who

5. Ibid., 22–27. Cf. Frank, *Nwn*, 5 (1917), 67; Kropp, *Zeitschrift für philosophische Forschung*, 6:3 (1952), 438–40; Heller, *Ernst Mach* (1964), 133–36; and Feuer, *Einstein* (1974), 341–45.

worried about fitting thought to experience with the fewest possible, and least abstract, concepts. Such barrenness is the sign of false prophets: "By their fruits ye shall know them!"[6] The main arguments of Planck's Leyden lecture recur, refined or amplified, in his speeches during the next twenty years and more. He regarded positivism as a present danger and, like any good campaigner, took as allies even people with whom, apart from the common cause, he disagreed about almost everything. He accepted Sir Oliver Lodge, who talked with spirits, for Lodge's "unshakable belief in a real outer world independent of us, in which mankind plays a relatively modest and subordinate role," and for his consequent rejection of "the positivist view, which rules out all questions related to this real outer world by declaring them to be meaningless."[7] Antipositivism is the leitmotiv of Planck's epistemological writings. It is therefore astonishing that some physicist-commentators imagine that he sang quite another tune. Léon Brillouin makes him a positivist; Albrecht Unsöld, in a tribute to him, construes the development of quantum theory as an anthropomorphizing of the world picture; Max Born uses the opportunity of the resultant confusion to deny that Planck's thought can be categorized at all.[8] The resistance or indifference of physicists to thinking systematically outside their discipline may help explain why Planck thought it important to state his message over and over again.

He first repeated himself in the spring of 1909, in his sec-

6. *PA*, *3*, 29 (1909). Cf. *PhW*, 105–13, which observes that the argument outlined in the last two paragraphs parallels Lenin's attack on Mach, also composed in 1908.

7. Planck to Lodge, 13 Dec. 1924 (Society for Psychical Research, London). Cf. *Wege*, v–vi (1933); *PhW*, 113; Goldberg, *HSPS*, 7 (1976), 151.

8. Brillouin, *Vie* (1959), 229; Unsöld, *Physik* (1958), 16, 29; Born, *PB*, *16* (1960), 147–48.

ond appearance as *philosophus in partibus*. The scene was Columbia University in New York City, where, although he learned nothing new scientifically, he discovered the vigor and hospitality of the New World and so (he wrote Wien) gained more than he had from any previous journey.[9] With his wits in very good order ("During the whole time [at Columbia] I drank not a drop of wine or beer, or came near any liquor, and consequently felt exceptionally well"),[10] Planck began his lectures in New York at the point where he had ended his diatribe in Leyden. The economical conception of physical theory might be plausible, clear, unambiguous, and antimetaphysical, he told his hospitable audience, "but—my friends—this view has never led to any progress in physics." He offered his historical evidence of the elimination of the anthropomorphic and the reason for it: "The goal is nothing other than the coherence and completeness of the system of theoretical physics, in fact the unity of the system not only in respect of all details, but also in respect of all physicists of all places, all times, all peoples, and all cultures."[11]

In September 1910, in a featured address to the Naturforscherversammlung in Königsberg, Planck took up the problem of unity again. His purpose now was not to save students from false prophets but to reassure physicists that the recent rapid revolution in their field, which had undercut the program of mechanical reduction, did not imply a postponement in the realization of the unified world picture. The theory of relativity and the principle of least action and, above

9. Planck to Wien, 9 Oct. 1912 (Autog. I/1365, SPK). Characteristically, Planck advised Wien, who planned a similar trip, to bring along a close relative (Planck took his daughter Emma): "It is an invaluable advantage to have a soul with you who shares all experiences and . . . considers all plans."

10. Planck to Runge, 28 July 1909 (RP).

11. Planck, *Acht Vorlesungen* (1910), "Vorwort," 4; cf. *Wege*, 179–200 (1929).

all, the universal constants had already identified themselves as building blocks of the new world picture; and relativity preeminently characterized the deanthropomorphizing tendency that marks every profound advance in physics.[12]

It need scarcely be said that a science independent of human aspirations did not agree with the feelings of Planck and his colleagues any better than it fit Machist solipsism. German-speaking theoretical physicists liked to put themselves forward as artists. It is not logic, Planck had written, but creative fancy, "which kindles the first flash of new knowledge in the mind of the researcher who pushes forward into dark regions"; without fancy, "good new ideas do not come."[13] He praised Hermann Minkowski's "artistically formed nature," Albert Einstein's special "power of imagination," and Arnold Sommerfeld's "forward-groping fancy."[14] Einstein in return remarked Planck's "truly artistic style" and the "artistic compulsion" that drove his creativity.[15] How can ideas arising from the peculiar genius of individual human beings add up to a deanthropomorphic world picture? Planck answered in the rhetoric of relativity: by seeking the "invariants" of the systems of physics. This is the duty of the physicist, and his faith. "If any thought strengthens and elevates us during the patient and often modest detailed work that demands our minds and bodies, it is that in physics we labor not for the day, not for momentary success, but, as it were, for eternity."[16]

Planck's attack on Mach was variously received. It is prob-

12. *PA*, *3*, 31, 44–45 (1910); cf. *Wege*, 158 (1925); *PhW*, 157–58.

13. Respectively, Planck, *Thermochemie* (1893), iii, 30–31, and Planck to Leo Königsberger, 2 June 1899 (acc. Darmst. 1922.93, SPK).

14. *PA*, *3*, 45 (1910), re Minkowski; *PAA*, 23 (1914), re Einstein; *PA*, *3*, 370 (1938), re Sommerfeld.

15. Einstein, *Nwn*, *1* (1913), 1079. Cf. Wien, *Ziele* (1914); Stark, *Prinzipien der Atomdynamik*, *1* (1910), 2; Forman, in Stehr and Maja, eds., *Society and knowledge* (1984), 342–43.

16. Planck, *Acht Vorlesungen* (1910), 126–27.

able that his audiences in Leyden and New York, where the central-European style of positivism had few adherents, applauded his commonsense realism. Most published commentary, however, came from Germans or Austrians who had already identified their epistemologies with Mach's. Wilhelm Ostwald chided Planck for painting the world picture with colors suitable only for physics. How then explain psychological phenomena? It was precisely Mach's point, according to Ostwald, that a scientist should be able to move from a description of the physical to a description of the psychological without changing his principles. Only a being endowed with the peculiarly human could take such a step. A truly unified science cannot be attained in the manner proposed by the would-be unifier of the world picture.

The same point was made by the old Machist Joseph Petzold, who noted that unification was Mach's strength: "Instead of attacking Mach, Planck should therefore have had every incentive to appeal to him as a predecessor."[17] Einstein's old friend Friedrich Adler echoed Petzold's complaint that Planck had much in common with Mach until his final leap of faith, which carried him to his concept of unchanging reality.[18] Einstein himself took the same line in a letter to Mach: "You have had so great an influence on the epistemological ideas of the younger generation of physicists that even your current opponents, like Planck, would doubtless be considered Machists by any physicist active a few decades ago, when all physicists were Machists."[19]

Mach did not care to blur the differences between Planck's

17. Ostwald, *Annalen der Naturphilosophie*, *10* (1910), 105–6; Petzold, *Zeitschrift für allgemeine Physiologie*, *10* (1910), 107–12. Cf. the even-handed Gerhards, *VWPS*, *36* (1912), 29.

18. Cf. Blackmore, *Ernst Mach* (1972), 222; Kleinpeter, *VWPS, 36* (1912), 5; and Merten, *Nwn*, *25* (1937), 253–4.

19. Einstein to Mach, 9 Aug. 1909, in Herneck, *Forschungen und Fortschritte*, *37* (1963), 241. Cf. Blackmore, *Ernst Mach* (1972), 223–26.

views and his own. In a sharply worded reply, he reviewed the elements of his epistemology, reaffirming that science had its origins in biological necessity and that scientific concepts, like natural species, survive only if "economical," that is, if they are convenient and reliable shorthands for multitudinous experience.[20] There can be no test of the truth of science by comparison with a real world independent of the observer. "In general, in every observation, in every opinion, both the environment and the observer enter into the expression. . . . Concern for a physics valid for all times and all peoples including Martians seems to me very premature, even almost comic, while many everyday physical questions press upon us."[21]

Mach then went to work on Planck's religious metaphors. "If belief in the reality of atoms is so essential to you, I will have nothing more to do with physical thinking, I will not be a proper physicist, I renounce all scientific reputation, in a word, no, thank you, to the community of believers. Freedom of thought is more precious to me." As far as Mach was concerned, Planck's religion of atoms, like all other religions, rested on prejudice and ignorance: in Planck's case, a prejudice that based the world picture on physics alone and an ignorance that proselytized without an adequate knowledge of the epistemological issues at stake.[22]

Planck did not care to be "badgered" (as he described his treatment) with religious metaphors or dismissed as a philosophaster. He decided to strike against Mach as a physicist as well as an epistemologist. In July 1910 he wrote Max von Laue and Wilhelm Wien in almost the same words: "Until October I will allow him his pleasure in the article; later he will wish that he had never written it. . . . Naturally I should like to avoid everything that could wound the worthy

20. Mach, *Physikalische Zeitschrift, 11* (1910), 600.
21. Ibid., 602, 603.
22. Ibid., 603, 606.

old man personally, but now I must rise up against his 'antimetaphysical' theory; I owe that to my own convictions."[23] It is hard to see how Planck could have believed that his answer would not wound Mach. He began by explaining his turning away from Mach's philosophy as a consequence of discovering its radical falsity: "The elimination of all metaphysical elements from the epistemology of physics cannot be accomplished in any way." Having failed to fulfill its impossible promise, Machism held nothing for Planck but a formalistic value, that is, no value at all, "because the first and foremost characteristic of all scientific research, the development of a constant world picture, is foreign to it."[24]

Planck proceeded to judge Mach's work by its fruits. Picking from the tree closest to home, he showed the inadequacy of Mach's grasp of the principles of thermodynamics and temperature, as set forth in Mach's *Principles of the theory of heat*; and he impugned the famous *Science of mechanics* as devoid of any useful result, theorem, or fruitful suggestion deriving from the biological-economical view of science. "It is exactly the opposite: where Mach tries to proceed independently in the sense of his epistemology [as in his discussion of the relativity of rotational motion], he very often ends in error." The physicist wishing to advance his discipline must be a realist, not an economist. "In the play of the phenomena he must search above all for the enduring, the indestructible, for what is independent of human senses. . . . That is as it always was and always will be, despite E. Mach and his supposed antimetaphysics."[25]

This time Planck's inner compulsion almost pushed him over the limits then allowable in philosophical combat. Mach's followers protested against Planck's effort to dis-

23. Planck to Laue, 4 July and 5 Aug. 1910, in Thiele, *Centaurus* (1968), 90; Planck to Wien, 6 July 1910 (AHQP).
24. Planck, *Physikalische Zeitschrift, 11* (1910), 1187–88.
25. Ibid., 1189–90.

credit their leader: it was "painful and offensive," "an unprecedented irresponsibility of judging and sentencing by a
mature scientist."[26] Einstein wondered how Planck could
have misunderstood Mach's purposes so badly as to think
them inimical to science.[27] At the time Einstein still regarded
Mach as his guide and his own identification of inertial with
gravitational mass as a brilliant confirmation of Mach's *Mechanics*, "despite Planck's unjustified criticism."[28] But soon
the leading German theorists sustained Planck.

In an obituary of Mach published in 1917, Sommerfeld
dismissed the late philosopher's epistemology as neither
healthy nor fruitful, and he observed that Einstein could not
have reached relativity from the principle of economy. "The
groping scientist, striving toward a hazy goal along a dark
way, needs a brighter lodestar than Mach's theory."[29] At
about the same time Einstein began to talk like Planck, but
in his own idiom. "[Mach's hobbyhorse] can give birth to
nothing living, but only stamp out harmful worms." He first
made public his mature view of Mach's teachings in 1922,
after having learned from the preface to the second, post-
humous edition of Mach's *Principles of optics* that his intel-

26. Respectively, Carl Cranz, Technische Hochschule, Berlin, to Mach,
4 Oct. 1910, and Paul Jensen, physiologist, Göttingen Univ., to Mach, 4
Apr. 1911, in Thiele, *Centaurus*, *13* (1968), 88, 90. Cf. Sommerfeld,
Nwn, *5* (1918), 199: the only time Planck was known to have acted "out
of his matter-of-fact reserve . . . [was] in his almost passionate polemic"
with Mach.

27. Einstein to Mach, ca. 1912, in Thiele, *Centaurus*, *13* (1968), 86;
Herneck, *Forschungen und Fortschritte*, *37* (1963), 241–42. Einstein's answer, in Planck, *New science* (1959), xi: Planck "probably felt that Mach
did not appreciate the physicist's longing for perception of this pre-estab-
lished harmony" between experience and theory.

28. Einstein to Mach, 25 June 1913, in Thiele, *Centaurus*, *13* (1968),
86.

29. Sommerfeld, "Ernst Mach," AW, Munich, *Jahrbuch*, 1917, 65–66.

lectual hero had repudiated relativity as dogmatic and un-
economical. He then recognized that what he had found
strengthening about Mach was his fierce independence, not
his extreme positivism. "That is a poor point of view. . . .
Insofar as he was a good student of mechanics, he was a de-
plorable philosopher."[30] And, to round out the group, Laue
came to stress that theoretical physics arises from a deep-felt
need to perfect "the physical world picture in the sense of its
unity."[31]

Anyone who contributes so importantly as Mach did to
the formation of an Einstein or a Planck need not hesitate to
be judged by the fruits of his teachings. Planck's requirement that a good epistemology should produce new physical
knowledge as well as create a mood conducive to production
appears overly strict.[32] The arguments of Planck and Mach
did not run on the same level. The one pushed commonsense realism as the only philosophy that caught his own
purpose and creative experience. The other taught an extreme positivism as the only system that made sense of the
origin and evolution of scientific ideas in time and place, and
of the empirical mode of acquisition of scientific knowledge.[33]

In giving notice that the progressively more abstract
world picture asymptotically approaches the real world,

30. Einstein to Besso, 13 May 1917, in Einstein and Besso, *Correspondance* (1972), 114; Einstein, discussion of relativity at the Société
française de philosophie, spring 1922, quoted by Herneck, *PB*, *15* (1959),
563–64; Holton, *Thematic origins* (1973), 226–31; Blackmore, *Ernst
Mach* (1972), 248–59.

31. Laue, "Antrittsrede," 30 June 1921, in *PPA*, 37; Kirsten and Körber, *2*, 256.

32. Cf. Kleinpeter, *VWPS*, *36* (1912), 11, and Feyerabend, *Studies in
history and philosophy of science*, *15* (1984), 1–22.

33. Cf. Gerhards, *VWPS*, *36* (1912), 24, 41, 45, 63; Frank, *Nwn*, *5*
(1917), 66–69.

Planck went against the concept of physics most frequently set forth by his colleagues around the turn of the century. According to most spokesmen then, physics aimed not at the truth, at a final account of things as they are, but only at an exact description of phenomena. The locus classicus of this "descriptionism" (to label it after a Victorian word for a mere describer) was the dictum of Gustav Kirchhoff, who in 1875 astonished his colleagues by declaring that analytical mechanics, the exemplar of exact physics, aimed not at truths of nature but at descriptions of phenomena. By 1881 Planck had already impugned this conception as allowing no place for the nobility of science or for its ability to treat matters not under observation, that is, to predict. "If the natural sciences could not predict, they would be at best no more than an amusement for scholars."[34]

With the agitation of Mach and the rise of Maxwell's electrodynamics and slipshod illustrative models to parity with mechanics, Kirchhoff's message gained acceptance; in 1900 descriptionism was the keynote of the grand International Congress of Physics convened in Paris to review and celebrate the triumphs of the nineteenth century. The same teaching—that the purpose of the scientist is to describe how the world affects his senses, not to discover the blueprints of nature—informed many contemporary addresses to learned societies, associations for the advancement of science, and other venues where scientists met their public. Among the causes of this unusual self-abnegation was the wish to defuse the attack then diffusely made upon natural science as an agent of spiritual and moral degeneracy, as an enemy of faith and family, as a menace to good literature, as an ally of socialism. By denying that they aimed at more than description, physicists withdrew from claims about the nature of

34. Planck to Runge, 28 Nov. 1881 (RP).

things that might bring them into conflict with defenders of established religion and moral order.[35]

The utility of Mach's philosophy as a weapon to defend "the structure of physics from attacks from outside" was understood by perceptive observers.[36] Although Planck stood firmly for, and was coming to represent, the high cultural and intellectual value of science, he too recognized the need for a public statement denying that science had a monopoly on the truth. In his Columbia lectures he referred to the dangers of under- and overvaluing theoretical physics, and he picked overvaluation as the more threatening. People not acquainted with the subject, he said, think that we will soon be able "not only to penetrate fully into the inner constitution of atoms but also to control the laws of spiritual life with physical formulas." He warned that there was no evidence in favor of either expectation, and much against both. In a public speech at the Berlin Academy in 1913, he attacked people who claimed to be able to construct a satisfactory worldview from science alone. That would be frivolous, he said, and an injury to both science and spiritual values. His method of resolving their opposition was not to relax the truth claims of science, in the manner of descriptionism, but to allow other claims and to attempt to unify the whole.[37]

The exchange with Mach gave Planck the status of a philosopher. To him it did not represent a promotion: philosophy, he said, was arbitrary, and every man was entitled to choose his own; science was "obligatory," the same for all mankind, and more important in proportion to its larger constituency. But his new status promoted him in others' opinion and opened a wider field of activity. "People com-

35. Heilbron, in Bernhard et al., eds., *Science* (1982), 51–73.

36. Frank, *Nwn*, 5 (1917), 67; cf. Heilbron, in Bernhard et al., eds., *Science* (1982), 59.

37. Planck, *Acht Vorlesungen* (1910), 2, and *Sb*, 1913, 75.

plain," wrote theologian Harnack, "that our generation has no philosopher. Unjustly: they now belong to other faculties. Their names are Max Planck and Albert Einstein."[38]

In the Academy

For Planck the unity of science connoted not only a common conceptual basis but also a corresponding institutional expression. What generalized mechanics was to the one so was the academy to the other. General-purpose learned societies like the Berlin Academy shared, in rhetoric at least, the goal of comprehending all the human sciences and the fiction that members, despite disciplinary differences, pursued common objectives in a common manner. The leading German academies heightened the integration of the knowers, if not of the known, by forming a cartel to support scientific expeditions and editions of classical works; Planck had the agreeable assignment, as Berlin's delegate to the cartel, of trying to arrange for publication of Boltzmann's collected scientific papers.[39] But even the cartel did not represent fully the unification and the politics of knowledge. Above it reigned the International Association of Academies, founded in 1899 in part to combat the growing influence and centrifugal force of professional organizations such as the Physical Society. It had as its highest aspiration an international, objective, unified, cooperative march toward Truth.[40]

38. Respectively, Planck to Harnack, 1914, in Kangro, *Vorgeschichte* (1970), 227, and Harnack, 1911, as quoted in Seelig, *Albert Einstein* (1952), 45.

39. Planck to Sommerfeld, 18 Jan., 7 and 11 Mar. 1908 (AHQP/84). Sommerfeld refused the cartel's invitation to edit Boltzmann; Planck turned to Boltzmann's successor, Fritz Hasenöhrl, who completed the work in 1909.

40. Schroeder-Gudehus, *Scientifiques* (1978), 42–48.

Planck shared this aspiration, which he expressed most fully when it was especially difficult to honor, just after the Great War. He then set forth the academic ideal in a memorial to his old friend Heinrich Rubens:

Rubens was not only a physicist and research scientist, he was above all an academician. Among the many distinctions he received at home and abroad, he always held his election to the Berlin Academy as the highest honor. The reason was deeply grounded in his conception of the importance of this body, in his enthusiasm for the academic idea—that however many specialties science may split into, it remains fundamentally an indivisible whole. Therefore the Academy, which embodies this unity, not only has a representative character but also must serve the urgent practical task of furthering the association of individual sciences in common work.[41]

In 1912 Planck took on official responsibility for the academic ideal. He was elected one of the two standing secretaries of the Berlin Academy's Mathematical-Physical Class, one of the most influential positions in German science. His colleagues, knowing his devotion to their corporate ideal and his slavishness to duty, judged him to be by far the best equipped among them for the job. They gave him nineteen votes out of twenty, the odd one going to (and cast by?) Walther Nernst. Planck succeeded the astronomer Arthur

41. *PA*, *3*, 337–38 (1923); *PPA*, 73. Cf. Planck to Karl Schwarzschild, 19 May 1913 (AHQP), praising Schwarzschild's maiden speech at the Berlin Academy (quoted in Kirsten and Treder, *2*, 238–41), which argues that mathematics, astronomy, physics, and chemistry constitute "*one* science . . . , which like Greek culture can only be grasped as a whole"; and *Wege*, 243 (1933): "For science, objectively considered, forms an internally closed whole. Its division into fields is not based on the nature of things."

von Auwers, who had held office since 1878, and joined the anatomist Wilhelm Waldeyer, who served from 1896 to 1919. Their opposite numbers in the Philosophical-Historical Class were Hermann Diels (secretary 1895–1920), an embodiment of the academic ideal, who wrote on the philosophy, science, and technology of antiquity, and Gustav Roethe (1911–1926), a far-right philologist, whose chauvinism was to cause Planck trouble during the war. For the moment, however, the new secretary had a free hand. He soon made good his first important initiative by capturing Einstein for the Academy.

The Academy's four secretaries alternated every three months as its executive officer and acted collectively as its spokesman. They did much for little. For 1,800 marks, just twice the salary of an ordinary member, they set up and presided over meetings, oversaw the Academy's projects, ran its finances, and supervised the publications of its proceedings.[42] The last, which took place with astounding speed and efficiency, demanded continuous attention. Two days after handing in his manuscript a member expected to receive proofs, which came with a printer's boy who waited for their correction. A week later the member had reprints. Planck conducted his share of Academy business with old-fashioned manners and without a typewriter, anticipating contingencies, interpreting the rules liberally when necessary, keeping the peace. At leisure in the Academy's rooms, which were furnished like an elegant club, with plush chairs and newspapers, the members talked politics as well as science, and with increasing heat as war drew close. We are told that Planck always stayed cool and reasoned.[43]

42. LS, *1*, 228; Künzel, AW, ITGOW, *Kolloquien*, *35* (1984), 73.

43. Willstätter, *Aus meinem Leben* (1949), 231; Laue, *Nwn*, *35* (1948), 5–6. A rare example of Planck's bending, rather than amending, the rules appears in Planck to Emil Fischer, 11 Jan. 1918 (Fischer/28, TBL).

The demands for, as well as on, so exemplary an administrator of science ran high. The Physical Society also made use of his ability and sense of duty: he was often chosen a member of its governing board, and three times he served as its president. Characteristically, he favored concessions to members of the Society not resident in Berlin in order to ward off the establishment of a competing and consequently disunifying organization.[44]

In the last year before the war, when Planck added the rectorship of Berlin University to his service to science, Einstein publicly congratulated the faculty on its choice.[45] Planck took up his office on 15 October 1913 and found it interesting, burdensome, and fulfilling. To ease his burden, the Prussian education ministry established an *Extraordinariat* (associate professorship) in theoretical physics at the University. The position went first to Max Born, but so eager was Laue to return to Berlin, which he had left in 1909 to begin to mount the academic ladder, that he offered to exchange his full professorship at Frankfurt for Born's more junior position. The transaction was consummated in 1918. This establishment of a second chair in theoretical physics at the University (beyond the honorary professorship held by Einstein) was the lasting achievement of Planck's rectoral year.[46]

Planck opened his rectorate on 1 November 1913 by in-

44. Planck to Wien, 14 June 1914, answering a letter from Wien of 12 June favoring a split (AHQP).

45. Einstein, *Nwn*, *1* (1913), 1077. The inspiration for Einstein's article came from the publisher Ferdinand Springer, who wanted something from Einstein for his new journal, *Naturwissenschaften*; Davidis, *Wissenschaft und Buchhandel* (1985), 53.

46. Planck to Lorentz, 11 Oct. 1913, withdrawing from the Solvay congress of 1913 for lack of time (LTZ/4); Planck to Wien, 14 Dec. 1913 (AHQP); Planck to Paul Ehrenfest, 19 June 1914 (ESC 8/7). Born, in Einstein and Born, *Briefwechsel* (1969), 19, and in Born, *My life* (1978), 161, 164–65, 187.

troducing two American guest professors for their inaugural lectures. The event, attended by members of the imperial household, allowed the rector to indulge his sentiment for king and country by referring to ties between the Hohenzollern and the University and to press his academic ideal by recalling the vitality and optimism of the United States. It was just this confidence in the future and sense of purpose that, by implication, Germany lacked. Indeed, very close to home, Planck's own eldest son, Karl, then twenty-five, suffered from nervous depression and drifted from one job to another. In him Planck discerned "the general sickness of the disparity between wishing and doing" that was characteristic, he thought, of Karl's generation. He would soon have reason to revise his estimate.[47]

Planck's speeches during 1913 contrast pleasantly with the intoxicated, patriotic rhetoric to which many academicians rose during the centennial of the rising against Napoleon and the founding of the University of Berlin. Here is a taste of this language, from an address by Dietrich Schäfer, professor of history at the University and member of the Academy. It is not impossible, Schäfer said, that Germany again will face a fight for its existence. Should God will that the kaiser take to the field to protect the rights and honor of the fatherland, then today's academic youth will follow, crying, "Father! I glorify you!" In battle they will find glory, and also students from other countries. For, as Schäfer assured his audience in a fine paralogism, "science is by its nature international."[48] In response to such harangues, Planck advised the Academy that its business was peaceful. The

47. Planck, *Berliner akademische Nachrichten*, 8 (1913–14), 27 (1 Nov. 1913); Planck to Hans Hartmann, 7 Mar. 1912 (MPG), to Gustav Roethe, 29 July 1916 (Göttingen Univ. Library), and to Leo Graetz, 18 June 1888 (DM), re Karl.

48. Schäfer, "Festrede" (1913), 27, 29; cf. Benno Erdmann, *Monismus* (1914), 3–4.

academicians of 1813 had held their meetings as usual "and proved their patriotism in their own way, through calm and true fulfillment of duty, just as much as the young soldiers who fought in the field for the liberation of the fatherland."[49]

Although he then deprecated chauvinism, he saw a need to enhance national and communal feeling. Like many of his mandarin colleagues, he viewed with pessimism the intensifying trend away from the idealism supposed to underpin German culture and community. He worried about the growing divisiveness brought about by special interest groups, by the selfishness of individuals who did not recognize or would not do their duty. In January 1913, in his first public address as secretary of the Academy, he referred to certain showy, superficial elements that had usurped the places of "many of the nobler forces of the nation."[50] What he had in mind appears from his masterful rectoral address delivered on the eve of the outbreak of the world war:[51]

We do not know what tomorrow will bring; we only suspect that something great, along with something monstrous, will soon confront our people, that it will touch the life and property, the honor, and perhaps the existence of the nation. But we also see and feel how, in the fearful seriousness of the situation, everything that the country could call its own in physical and moral power came together with the speed of lightning and ignited a flame of holy wrath blazing to the heavens, while so much that had been considered important and desirable fell to the side, unnoticed, as worthless frippery.

Planck identified as menacing, disunifying elements people who did not look forward to a future faithful to, but not lim-

49. Planck, *Sb* (1913), 74 (23 Jan.).
50. Ibid.
51. *PA*, *3*, 77 (3 Aug. 1914).

ited by, the past; people who repudiated everything new and those who chased after every novelty; and people who fomented discord by multiplying parties, lobbies, and unions. It is the same in science, he said: woe to the investigator so in love with his new idea that he neglects to test it rigorously against received wisdom; woe also to the investigator so in love with his old conceptions that he refuses to weigh the merits of a new insight.

Planck ended this, the last of his prewar speeches, with a consideration that he had adumbrated at the outset of his career and that was to become a dominant theme in his writing: "The assumption of an absolute determinism is the essential foundation of every scientific inquiry."[52] From this dictum he did not except the problem of human will and ethical behavior; it is a fundamental error, according to Planck, to infer from our inner experience of freedom a lapse of strict determinism in our affairs. We can often predict reliably the future behavior of a person we know well, and we can perceive the chain of causes behind our own past actions. The resolution of the apparent paradox: studying a decision in the course of making it so disturbs the phenomenon under investigation as to make the outcome scientifically valueless. Science necessarily proffers a man no guidance in the greatest question of his life: how shall he behave? "He cannot find a complete answer to this question in determinism or causality or any pure science, but only in his moral conscience, his character, his worldview."[53]

And where does one acquire the worldview that prompts proper action? Here again Planck proposed to judge by fruits. The proper moral code is that which preserves society and maximizes opportunities for the intellectual and cultural development of individuals; the code, inculcated by liberal

52. *PA*, *3*, 89 (1914), anticipated in Planck to Runge, 28 Nov. 1881 (RP): the predictions that are the goal of science presuppose causality.
53. *PA*, *3*, 89 (1914).

Christianity, under which European civilization flourished; in a word, the honorable, instinctive behavior of men like Planck. This conception has much in common with Kant's critique of practical reason, which teaches among other things that moral value inheres in acts insofar as they derive from an immediate perception of duty and obligation. A few months before the outbreak of the Great War, Planck had occasion to endorse Kant's point of view.

A mathematician, Eduard Study, had sent Planck a copy of his book defending realism against the attacks of generations of idealists. Planck returned thanks and praise to Study for resting his defense on the practical value of realism, which, he wrote, agreed entirely with the spirit of Kant's practical reason. Kant was still the best guide. To be sure, his notions of space and time had been flawed by his ignorance of relativity and non-Euclidean geometry; but he had the foundations of knowledge right, just as Christ's preaching contained the elements of all true religions. "Farsighted theologians are now working to mine the eternal metal from the teachings of Jesus and to forge it for all time"; correspondingly, "it would be of incalculable importance for the development of the realistic worldview . . . if Kant's philosophy, which once again has shown itself to be a historical force, came completely into its own."[54] Planck would soon learn that even the powerful combination of the historically certified ethical values of Christianity and the Kantian imperatives of his class did not always make the course of action clear.

Rector of the University, secretary of the Academy, and factotum of the Physical Society—all that would fill a normal man's day, duty, and ambition. Not Planck. During his rectoral year he took up with still another organization, which was, in time, to give him his highest and widest platform. It

54. Planck to Study, 21 Dec. 1913 (Autog. I/383, SPK).

was the Kaiser-Wilhelm-Gesellschaft, founded in 1911 to create research institutes at private expense. The Society consisted of members who, for a substantial contribution, had the right to wear special dress and to breakfast annually with the emperor; a Senate, the maker of overall policy, drawn primarily from the membership; an executive committee (*Verwaltungsausschuss*), coopted from the Senate; and a president, the head of the executive committee, chosen by the Senate. From 1911 to 1930 the president of the Society was its main architect, Adolf von Harnack. According to Harnack's original plan, the Prussian government would provide land for research institutes, which would be built as funds and directors became available, and the Society would put up the structures and pay the operating expenses. Additionally, the government might set up positions within the institutes, whose incumbents would ipso facto become state employees.[55]

Planck was drawn into the Society's affairs in 1913, during negotiations to set up a Kaiser-Wilhelm Institute for Physics under Einstein's directorship. The Society itself and an independent fund for the support of German science, the Leopold-Koppel Foundation, offered to cover all outlays except a third of the annual operating expenses, for which Planck and his colleagues—Fritz Haber, Walther Nernst, Heinrich Rubens, and Emil Warburg—turned to the Prussian Kultusministerium, where they received a favorable response. The generous arrangement, by which the Koppel Foundation provided a building and it and the Society each pledged 25,000 marks a year for operating expenses, foundered when the finance minister declined to follow the positive recommendation of his colleague for culture.[56]

55. Burchardt, *Wissenschaftspolitik* (1975), 85–94, esp. 93; Wendel, *Kaiser-Wilhelm-Gesellschaft* (1975), 75–115.

56. Haber et al., "Antrag," Feb. 1914, and Harnack to Schmidt-Ott (Kultusminister), 12 Sep. 1917, in Kirsten and Treder, *1*, 146–49; Bur-

In his memorandum to the kaiser proposing the foundation of the Kaiser-Wilhelm-Gesellschaft, Harnack had emphasized that fundamental research was necessary to industrial advance and that "military power and science are the twin pillars of Germany's greatness." The Prussian finance minister could not make out how the particular science of pure physics, in an institute headed by Einstein, would assist the state in either an industrial or a shooting war; and he was undoubtedly not encouraged by the projectors' reference to the scholarly *Monumenta germaniæ historica*, a series on German medieval history, as an indication of the sort and level of support they desired.[57] The outbreak of war forced a pause in the fight of pure science for state support before Planck and his colleagues could outflank the minister.

In the War

Planck rejoiced in the patriotic surge of the fall of 1914. Although the father of two boys of military age and the rector of a university emptied by the call to the colors, he welcomed the sacrifice of self to the state as a most desirable unification of purpose and a consolidation of the national will. In this he was at one with his colleagues. As Harnack put it, the war brought "one will, one force, a holy seriousness of purpose"; it raised the individual "above common egoism, the egoism of parties, and all false goods and pseudo-ideals of a moral, aesthetic, and intellectual character"; it directed the "holy flame of the fatherland" against "everything selfish, petty, and common."[58]

chardt, *Wissenschaftspolitik* (1975), 118–19; Wendel, *Kaiser-Wilhelm-Gesellschaft* (1975), 197–201; KWG, Senat, "Protokoll," 21 Mar. 1914, 4–5, and 22 Oct. 1914, 1 (Fischer/1, TBL).

57. Harnack, in MPG, *50 Jahre* (1961), 83, 89 (1909); Kirsten and Treder, *1*, 147.

58. Harnack, *Schwelle* (1916), 3–4, and *Erforschtes* (1923), 279 (1916).

The enthusiasm of these first weeks, together with the firm conviction that Germany was engaged in a defensive war against unscrupulous opponents, carried Planck to a step that he soon regretted bitterly. He signed the "Appeal to the Cultured Peoples of the World" ("An die Kulturwelt! Ein Aufruf"), otherwise known as the "Appeal of the Ninety-three Intellectuals." This proclamation, issued on 4 October 1914 in all the principal German newspapers and in ten languages, declared the leaders of German art and science to be at one with the German army and repudiated the charges of the Entente powers that German forces had committed atrocities in Belgium. Drafted by Ludwig Fulda, a playwright who had been concerned with propagating German culture abroad, and polished by the mayor of Berlin, it traded on resistance to the expressed policy of the British to separate the German people from Prussian militarism and on the need widely felt in Germany to expose the "lies" of the Entente powers to the neutral states. Responding to these imperatives, many people, including Planck, signed the Appeal before they had had a chance to read it, on the strength of the reputations of the other signatories, who had lent their names on the same basis.[59]

Most irritating to these hasty intellectuals—more irritating even than the accusation of brutality—was the slander, as they saw it, that their troops had willfully and unnecessarily destroyed treasures of art and science. A country that had given birth to a Beethoven and a Goethe, they said in a grand non sequitur, knew how to respect the cultural heritage of Europe.[60] The professors could not believe that their troops,

59. Einstein to Lorentz, 2 Aug. 1915, in Nathan and Norden, *Einstein on peace* (1981), 11, on Planck's ignorance of the text; Wehberg, *Wider den Aufruf* (1920); vom Brocke, in Calder et al., eds., *Wilamowitz* (1985), 654–55, 692–93, 702–3, 708.

60. Planck's close colleagues Harnack and Nernst also signed the Appeal, the text of which has often been reprinted, for example, in Schroe-

led by boys they had educated, and commanded by men with whom they themselves had studied, had burned the library of Louvain. The same message was sent west in a proclamation by 3,016 university-level teachers, who pointed to the outstanding value of military service for all peacetime vocations, even the cultivation of science (it inculcates self-sacrifice, the observance of duty, respect for others), and in a declaration by the rectors of twenty-two universities, who pointed to the immunity to hoodlumism acquired in their institutions (they impress nothing so deeply on their students as "attention and wonder before the great creations of the human spirit in art, science, and technology, irrespective of their country or people").[61]

Planck made his new glorification of unity the theme of the business report of his rectoral year. The calling up and enlistment of young men, both students and instructors, had depopulated the University. That was not to be deplored. "The German people has found itself again." No one knew when things would return to normal. That did not matter. "One thing only we know, that we members of our university . . . will stand together as one man and hold fast until—despite the slander of our enemies—the entire world comes to recognize the truth and German honor." Writing to Wien just after leaving the rectorate, Planck found more to appre-

der-Gudehus, *Deutsche Wissenschaft* (1966), 272–74. For the chauvinism of the Berlin professors and students in the fall of 1914, see Basler, in Berlin, Univ., *Wissenschaftliche Zeitschrift*, 10 (1961), 182–84; for the chauvinism of the Academy, thirty-three of whose then current or later members signed the Appeal, see LS, 1, 175–76. Vom Brocke in Calder et al., eds., *Wilamowitz* (1985), 657–59, makes the point that the signers, many of whom had important political or administrative posts, were not starry-eyed intellectuals.

61. Respectively, "Erklärung der Hochschullehrer des deutschen Reiches" (16 Oct. 1914) and "Die Universitäten des deutschen Reiches an die Universitäten des Auslandes" (Sept. 1914), in *Berliner akademische Nachrichten*, 9 (1914–15), 35, 20.

ciate in the war than to regret. "Besides much that is horrible, there is also much that is unexpectedly great and beautiful: the smooth solution of the most difficult domestic political questions by the unification of all parties . . . , the extolling of everything good and noble."[62] There was already much that was horrible among Planck's close colleagues. Nernst had lost a son, and others would soon; Planck's children were all in danger, the twin girls, quickly trained by the Red Cross, awaiting assignment to hospitals, Karl at artillery school, Erwin at the front. "And yet [their father wrote their aunt] what a glorious time we are living in. It is a great feeling to be able to call oneself a German."[63]

In his speeches to the Academy, Planck continued to extol the spirit of October 1914 and to suppose it to persist.[64] But his belief in a united and common will was as forlorn as his hope that classical physics might be reconciled with the quantum of action; and by 1915 Planck found himself unable to agree with some of his closest friends. Wien had decided that for his generation the old relations among nations would never again be possible. As his contribution to the realization of perpetual enmity, he drew up a demand that physicists have nothing to do with British journals, except perhaps to publish in them replies to personal attacks; he claimed that British physicists appropriated discoveries made in Germany, confused truth and falsehood, argued in

62. Quoted in Basler, in Berlin, Univ., *Wissenschaftliche Zeitschrift*, *10* (1961), 182; Planck to Wien, 8 Nov. 1914 (AHQP). Cf. *PhW*, 220.

63. E. Fischer to T. W. Richards, 16 Nov. 1914, in Reingold and Reingold, eds., *Science in America* (1981), 229; Planck to Emma and Max Lenz, 17 Sept. 1914 (MPG).

64. For example, Leibniz Day, 1 July 1915, in *Sb* (1915), 483 (the collective spirit is a "suit of armor" to protect the most sensitive points of an individual's honor), and the kaiser's birthday, 25 Jan. 1917, in *Sb* (1917), 35–36 (the war has brought "the entire force of the nation to a single focus").

bad faith, and in general justified his conviction that England was the worst enemy of the Reich.[65]

Planck declined to sign Wien's manifesto, not because he disagreed with its content ("I have often enough felt just the same") but because he thought that it would be considered, as it was intended, as an act of war, an impotent gesture by frustrated professors. Such matters should be saved for peacetime, Planck told Wien, when, should Germany win the war, things would be composed as Wien wished. If, however, the unthinkable should happen and Germany should lose, the consequences of the manifesto might be very serious indeed.[66] Wien did not find this argument persuasive, and he issued his manifesto. It bore the signatures of sixteen physicists including Sommerfeld, Johannes Stark (of whom more later), and Wien himself.

Planck's move toward moderation accelerated under the influence of Lorentz, who was perfectly placed to act as a conscience. Lorentz was at home with all the languages and cultures of western Europe; he was a national of a neutral state; and he had won Planck's respect as a scientist and as a man. Their friendship dated from Planck's visit to Leyden and his attack on Mach. They met again at the Solvay council in Brussels, where Lorentz demonstrated that easy and natural command of sciences, languages, and people that made him the chairman of choice at international meetings. What struck Planck and also Einstein most forcefully about

65. "Aufforderung" (AHQP); Wien to C. W. Oseen, 19 Nov. 1914, and to August Sperl, 1 July 1915, in Wien, *Aus dem Leben* (1930), 60, 62. Cf. Lenard, *Quantitatives* (1918), 14–15; Badash, Royal Society of London, *Notes and records*, *34* (1979), 108, gives examples of British counterclaims.

66. Planck to Wien, 1 Jan. 1915 (AHQP). Cf. Forman, *Isis*, *64* (1973), 157; the faithfully drawn protagonist in McCormmach's *Night thoughts* (1982), 143–52, 210–14, held opinions similar to Planck's and refused to follow Wien.

Lorentz was "the admirable clear harmony of his entire personality—a true image of his relationship to his science and his fellow men."[67]

Lorentz's inner harmony and intellectual cosmopolitanism were sorely tried by the fate of Belgium and the several manifestos from the scientists on either side. He sent Planck accounts of the German occupation, which showed, as Planck put it, that many things had occurred "that do not conduce to the honor of Germans."[68] But the suffering was great, unheard of, on both sides, Planck wrote: his nephew, a physicist, the only son of his brother, had been killed, his own son Erwin had been taken prisoner, both sons of Nernst had perished in battle. "Where is the compensation for all this unspeakable suffering?" Planck could find it only in the spirit of the first weeks of the war. "Could it really be such a terrible thing that has called forth so willing a sacrifice, so pious an enthusiasm? I cannot believe it."[69] During 1915 they exchanged visits, Lorentz to Berlin and Planck to Leyden, to review the claims and counterclaims. Planck brought Lorentz to see some things from the German point of view. Lorentz brought Planck to a courageous, public, quasi-repudiation of the Appeal of 1914.[70]

Already in March of 1915 Planck was apologizing in private for the manifesto. He had signed it only, he told Lorentz, out of a conviction that Germany was fighting a war for its existence and had to stand, and be seen to stand, united. That was all the signing meant, at least to Planck and Harnack; having declared their solidarity with their country-

67. *PA*, *3*, 179 (1929), 343 (1928, quote); cf. ibid., 347 (1928), 414 (1938); Einstein to Zangger, 16 Nov. 1911, in Seelig, *Helle Zeit* (1956), 43.

68. Planck to Lorentz, 15 Nov. 1914 and 8 Aug. 1915 (LTZ/5).

69. Planck to Lorentz, 28 Mar. 1915 (LTZ/5).

70. Planck to Wien, 4 May 1915 (AHQP); Planck to Lorentz, 27 Feb. 1916 (LTZ/5).

men, they could turn away from the home front and try to reach an understanding with their enemies. "In my opinion, scientists [now] face no more urgent or finer a challenge than to do their best quietly to counter the continuing poisoning of [the minds] of fighting men and the deepening of hatred among peoples."[71]

Planck had tried to put this policy into practice. He adhered to the Kulturbund deutscher Gelehrter und Künstler (Cultural Association of German Scholars and Artists), chaired by his fellow secretary Wilhelm Waldeyer. Its purpose, as Planck set it out for Svante Arrhenius a month after the appearance of the Appeal, was "the support of all efforts aimed at bringing out the truth." The Association was preparing for peace, when old scholars would have their turn at the front, helping to restore old ties. "We consider it our obligation to help in carefully cultivating what remains of international good will and to oppose as far as possible every cheapening and poisoning of public opinion by impassioned libel and slander." Unhappily the Association, which was promoted by the organizers of the Appeal of the Ninety-three Intellectuals, could not find the truth, let alone declare it; nor could it long retain the highly prized unity that accompanied and sanctified mobilization. The purpose of Planck's explanations to Arrhenius was to counter a highly political and pro-German statement made to a Swedish reporter by Ostwald, formerly the head of the pacifist Deutscher Monistenbund, who claimed to be speaking for the Association.[72]

Speaking for himself before the Academy in July 1915, and also for other "good Germans" who wanted nothing from the war but the defense of German honor and the res-

71. Planck to Lorentz, 28 Mar. 1915 (LTZ/5).

72. Planck to Arrhenius, 15 Nov. 1914 (Stockholm Univ. Library); LS, *1*, 176–77; vom Brocke, in Calder et al., eds., *Wilamowitz* (1985), 664, 700, 719.

toration of international relations, Planck hinted that the
Reich might bear some blame for the war. In entertaining
this possibility he diverged not only from Ostwald and Wien
but also from the more moderate Sommerfeld, with whom
he usually agreed about lesser matters like the nature of
physics.[73] Working down as usual from a great height, Planck
observed that since science outlasts all earthly things and
events, the war would in the end be the subject of objective
research and that it was as unlikely that one side would come
out all black and the other all white as that science should
arrive at ultimate truth.[74]

Planck made his moderation more widely known early in
1916, in the form of an open letter to Lorentz. On Lorentz's
advice, he softened the defensive portion of his draft statement to refer to the special circumstances of the first weeks
of the war, to drop a countercharge of slander, and to allow
for the possibility that Germany might bear some responsibility for the fighting. But the fine, final lines that place some
intellectual and moral values above identification with country are as Planck first wrote them.

The text that Lorentz had printed runs as follows:[75]

> The well-known "Appeal to the Cultured Peoples of
> the World," published in October 1914 over the signa-
> tures of ninety-three German scientists and artists, has
> by its {formulation, as I have repeatedly noticed with
> distress, given rise to incorrect ideas about the feelings

73. Planck to Arrhenius, 6 Dec. 1914 (Stockholm Univ. Library); *Sb*
(1915), 484; Benz, *Arnold Sommerfeld* (1975), 85.

74. Cf. Harnack, *Erforschtes* (1923), 33–34, and Schwabe, *Wissenschaft und Kriegsmoral* (1969), 195, n. 24.

75. Substantive differences from the original text, enclosed in Planck
to Lorentz, 27 Feb. 1915, are indicated by curly brackets; the corresponding original text, if any, is given in the Notes. The texts and a draft of
Lorentz's suggestions, dated 20 Mar. 1916, are in LTZ/5.

of its signers.}[76] In my personal view, which I know is shared in its essentials by many of my colleagues, including Adolf von Harnack, Walther Nernst, Wilhelm Waldeyer, and Ulrich von Wilamowitz-Möllendorf, the Appeal{, whose form reflects the patriotic fervor of the first weeks of the war,} could and should signify nothing but an act of defense: above all,[77] the defense of the German army against the bitter accusations raised against it, and an[78] emphatic affirmation that German scientists and artists would not separate their cause from the cause of the German army. The German army is nothing other than the German people in arms, and, like people of all callings, scientists and artists are inseparably bound to it.

That we do not defend the behavior of every German, either in peace or in war, I will not stress, although I hold it to be just as self-evident as the fact that we do not now {possess a definite scientific answer to the great questions of the historical present. Only a later, fully objective investigation, whose results we can look forward to with a clear conscience, can decide where the responsibility ultimately lies for the failure of the peace initiatives and for all the resulting human suffering.}[79]

As long as this war lasts we Germans have only one task, to serve the nation with all our strength. But what

76. Originally: "form unfortunately gave rise to many misinterpretations and to an altogether false conception of the feelings of its signers."

77. Here the original has additionally: "against the unprecedented campaign of slander that surprised the German people at the outbreak of the war (you will excuse me for not giving examples), but above all."

78. Here the original has additionally: "Thence arose the necessity of an explicit."

79. Originally: "already have a premature assessment of the great questions of the historical present."

I want to express to you with special force is the firm conviction, which even the events of this war cannot change, that there are domains of intellectual and moral life that lie beyond the struggles of nations, and that honorable cooperation in the cultivation of these international cultural values and, not less, personal respect for citizens of enemy states are indeed compatible with ardent love and energetic work for one's own country.[80]

Planck asked Lorentz to send copies of the printed letter to physicists in the Allied countries whose good opinion he valued: participants in the first Solvay council; W. H. and W. L. Bragg, Joseph Larmor, Oliver Lodge, Lord Rayleigh, Arthur Schuster, and J. J. Thomson in England; O. Chwolson and B. Galatzine in Russia; Pierre Duhem in France; and Vito Volterra in Italy. The main point, Planck wrote, was that his good will should come through; even though that might not mollify others, it would greatly relieve him. "But whatever the effect might be, good or bad, I will in any case have freed myself from a mental scruple that has weighed on me for a year and a half."[81]

The initial public response in the Entente countries was not favorable. In an incoherent but representative comment, the editor of the news magazine of British astronomy *The observatory*, which printed an English translation of Planck's letter, observed that it did not cancel the identification of the signatories with the acts of the German army, "so they accept responsibility for the sinking of the *Lusi-*

80. Cf. Planck's response to Einstein's inaugural address to the Academy, 2 July 1914 (*PPA*, 25; Kirsten and Körber, 2, 248): in physics it is "easier than in other sciences to settle the sharpest objective considerations in high personal esteem and in hearty friendly feeling."

81. Planck to Lorentz, 28 Mar. 1916; he again mentioned unburdening his conscience in letters of 28 Apr. and 23 Oct. 1916 (LTZ/5).

tania."[82] Planck could not withdraw other people's support of the Appeal, and he explicitly refused to defend every act of the German military. Eventually, however, his statement earned him credit among reasonable people on both sides of the conflict.

Planck's firm endorsement of transcendental and international values did not stop with words. He applied to the head of the German government in Belgium on behalf of Ernest Solvay and others to whose sufferings he was alerted by Lorentz.[83] At home he worked harder than ever to moderate academic passions. In June and July 1915, soon after the novel application of science to war in the gas attack at Ypres, the Academy's long-simmering relations with enemy savants and learned societies came to a boil. It had fought over whether to award prizes to non-German authors in its essay competitions. It had agitated whether to respond to French initiatives and expel from its body corresponding members holding citizenship in any of the Entente powers. It had perplexed itself over the diatribes of Sir William Ramsay and other rabid elements of the Royal Society of London. Planck led the opposition against shortsighted policies that might damage irretrievably that precious European heritage, "the institution of academies."

Did anyone doubt that the international cooperation represented by the Academy should be reestablished after the war? Planck's reply: "Whoever could so much as hesitate to answer this question affirmatively would sin against the spirit of science." He drew up a careful case against several propositions for purges circulating in the Academy. At the showdown on 22 July 1915, his proposal to delay all deci-

sions about academies of other nations until the end of the
war passed by a majority of two to one. He then had a hard
fight to defeat a rider that specified the decision: breaking
off all relations with corresponding members, academies,
and learned societies in enemy states after the peace. As Einstein observed, academicians were not notable for their "independence of character, freedom from the prejudices of
caste, and spirit of self-sacrifice." But the Academy did eventually listen to Planck, and in 1916 he could write to a fellow
secretary that he was truly sorry to learn of Ramsay's death,
since, owing to the relative moderation of the Germans,
Ramsay's "undignified behavior toward German science
could only have led to an atonement in favor of the Academy
after the coming of peace."[84]

Similarly, Planck opposed his rabid colleagues at the University who supported the expansionist policy of the right-
wing Alldeutscher Verband (Pan-German League). In June
1915 a committee dominated by Schäfer and promoted by
the League trotted out an "intellectuals' memorandum" at a
formal ceremony in Berlin. Some 1,347 persons signed, including 352 teachers, 50 of whom taught at the University
of Berlin. A "counter declaration," drawn up chiefly by Hans
Delbrück, "the best type of liberal German scholar," received only 141 signatures, but 15 from Berlin professors including Planck, Einstein, and Harnack. Delbrück's declaration plumped for defensible borders but opposed securing
them by annexation. It also urged internal reform.[85] What

84. Basler, in Berlin, Univ., *Wissenschaftliche Zeitschrift*, *10* (1961),
189–90; Planck, *Sb*, 1915 (Leibniz Day, 1 July), 483–84; LS, *1*, 184–86,
227–28; Einstein, *Sozialistische Monatshefte*, no. 25–26 (1919), 1055–
56; Planck to Roethe, 29 July 1916 (Göttingen Univ. Library). On corresponding moves in Britain, where J. J. Thomson as president of the
Royal Society played a part similar to Planck's, see Badash, Royal Society
of London, *Notes and records*, *34* (1979), 94–97, 110–12.

85. Basler, Berlin, Univ., *Wissenschaftliche Zeitschrift*, *10* (1961),
186–89; LS, *1*, 178–79; the praise of Delbrück comes from Born, *My life*

The doctor of philosophy, 1879. Courtesy of Niels Bohr Library, American Institute of Physics, New York.

The professor in his library. Copyright Ullstein Bilderdienst; courtesy of Max-Planck-Gesellschaft, Berlin.

The lecturer relaxing during a speaking engagement in Bonn, 1928.
Courtesy of Max-Planck-Gesellschaft, Berlin.

The son and grandson of pastors, 1935. Courtesy of Niels Bohr
Library, American Institute of Physics, New York.

The musician seeking solace, 1942. Courtesy of Max-Planck-Gesellschaft, Berlin.

Planck at work under portraits of his first wife, née Marie Merck, and their four children. Courtesy of Niels Bohr Library, American Institute of Physics, New York.

The elder son, Karl, between his sisters Emma and Grete, ca. 1900.
Courtesy of Max-Planck-Gesellschaft, Berlin.

Erwin Planck, younger son of Max and Marie Planck, trying to orient himself and his father in the mountains. Courtesy of Max-Planck-Gesellschaft, Berlin.

Planck and his second wife, née Marga von Hoesslin, 1931. Copyright Bilderdienst Süddeutscher Verlag; courtesy of Niels Bohr Library, American Institute of Physics, New York.

Planck with (from the left) Walther Nernst, Albert Einstein, Robert A. Millikan, Max von Laue, 1928. Copyright Ullstein Bilderdienst; courtesy of Max-Planck-Gesellschaft, Berlin.

The four secretaries of the Berlin Academy of Sciences, ca. 1930: (from the left) Heinrich Lüders, Ernst Heymann (who took action on the part of the Academy against Einstein), Max Planck, and Max Rubner. Copyright Ullstein Bilderdienst; courtesy of Max-Planck-Gesellschaft, Berlin.

The award of the Planck medal to Einstein, 1929. Courtesy of Niels Bohr Library, American Institute of Physics, New York.

Planck with Niels Bohr in Copenhagen, 1930. Planck's head is encircled by what appears to be a Bohr orbit. Copyright Niels Bohr Institut, Copenhagen; courtesy of Niels Bohr Library, American Institute of Physics, New York.

Planck loaded with the honors of office, 1933. The man on the
right, from whom Planck has averted his head, is Wilhelm Frick,
Nazi minister of the interior, who designed the law that dismissed
Jews from the civil service. Courtesy of Max-Planck-Gesellschaft,
Berlin.

Planck with Max von Laue,
1947. Courtesy of Max-
Planck-Gesellschaft, Berlin.

The original buildings of the Physikalisch-Technische Reichs-anstalt. Copyright Physikalisch-Technische Bundesanstalt; courtesy of Max-Planck-Gesellschaft, Berlin.

Berliner-Schloss, administrative seat of the Kaiser-Wilhelm-Gesellschaft. Courtesy of Max-Planck-Gesellschaft, Berlin.

Planck at the inauguration of the Kaiser-Wilhelm-Institut für Metallforschung, Stuttgart, 1935. Courtesy of Max-Planck-Gesellschaft, Berlin.

The office of the president of the Kaiser-Wilhelm-Gesellschaft. Courtesy of Max-Planck-Gesellschaft, Berlin.

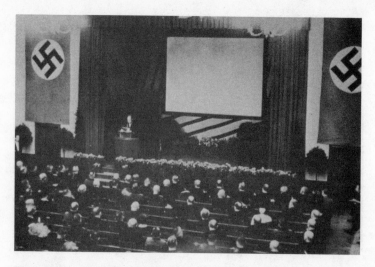

Planck addressing the 25th anniversary meeting of the Kaiser-Wilhelm-Gesellschaft, 1936. Courtesy of Max-Planck-Gesellschaft, Berlin.

The Kaiser-Wilhelm-Institut für Physik, ca. 1937; the cylindrical tower on the left housed the high-tension accelerator. Courtesy of Max-Planck-Gesellschaft, Berlin.

Planck wished or would tolerate in this line may perhaps be inferred from Harnack's views, which called for universal suffrage, complete religious freedom, the right of free association, and the deflation of the "caste spirit." All of this was to be initiated in a conservative manner, from the top down: "'sound progress' and 'conservative' are not only not contradictions, but inseparably bound together."[86]

Against Fate

Conservative reform proved as unattainable as victory. As the Reich staggered under looming defeat, wide-ranging changes forced from below seemed inevitable. The circumstances immobilized Planck. On the one hand, he had come to see the necessity for many of the measures advocated by the liberal, outlawed Bund Neues Deutschland (New Germany League), to which Einstein belonged. On the other hand, he could not take a step or sign a proclamation demanding the removal of the kaiser, as the League did publicly after its official reappearance in August 1918, although he recognized that abdication was the precondition to securing the reforms he believed necessary to the salvation of the state. He expressed his dilemma in a confidential letter to Einstein:[87]

And here is something else I believe I see clearly and would also gladly do my best to work toward: that it would be a great stroke of fortune for us, indeed a sav-

(1978), 167. Texts of Schäfer's and Delbrück's declarations are printed in Grumbach, *Das annexionistische Deutschland* (1917), 132–40 and 409–11, respectively.

86. Harnack, *Erforschtes* (1923), 287–88 (quote), 294–95 (1916), 298–307 (1917); Planck to Einstein, 29 Dec. 1917 (AE).

87. Planck to Einstein, 26 Oct. 1918 (AE); Clark, *Einstein* (1971), 188, 198; Kirsten and Treder, *1*, 213–14; Fricke, ed., *Die bürgerlichen Parteien* (1968), *1*, 179–82.

ing grace, if the bearer of the crown would voluntarily renounce his rights. But the word "voluntary" makes it impossible for me to come forward in the matter; for first I think of my sworn oath, and second, I feel something that you will not understand at all . . . , namely, piety and an unbreakable attachment to the state to which I belong and which is embodied in the person of the monarch.

Two days after Planck had thus expressed his political philosophy, the Reichstag declared a parliamentary government. During the next week, the sailors mutinied in Kiel; by 7 November the revolution they had started had spread to Hamburg and Bremen; workers' and soldiers' soviets appeared throughout Germany; on 9 November a general strike began in Berlin, and the kaiser abdicated without Planck's help. His act freed his liegeman's conscience. Planck gave his support to a new political party, the moderate-right Deutsche Volkspartei, which was preferred by many academics and industrialists. It promised to work within the new political order for economic and social reform.[88]

The breakup of Planck's external world coincided with a personal tragedy of Job-like proportions. Only one of his four children, his younger son Erwin, who had been a prisoner in France, long survived the war. The elder son, Karl, who had had no purpose, died of wounds suffered in action. Karl's sacrifice hit Planck hard. It was not the death itself; as he wrote Lorentz, "everyone should be happy and proud to be able to sacrifice something for the whole." Nor was it that the blow that had struck so many of his friends with sons

88. *PhW*, 224, citing a questionnaire of 1939; Planck to Max Schirmer, 17 May 1931, mentioning allegiance to the DVP (MPG); "Aufruf der Deutschen Volkspartei" (18 Dec. 1918), in Salomon, *Die deutschen Parteiprogramme*, *3*, (1920), 86–89, and the "Wahlaufruf" of the Bavarian DVP, ibid., 90–94. Cf. Forman, *Minerva*, *12* (1974), 49; LS, *2*, 242; Laue, *Nwn*, *45* (1958), 223.

of military age had come to him, although he now knew their pain. "The suffering that war brings really hurts only when one feels it in his own body." The grief lay in the discovery that he had undervalued his son. Before the war, when Karl drifted, unable to settle into one of the high professions available to and expected of him, his life did not seem so precious to his father as its end showed it to be. The call to the colors had ennobled Karl. In words that might stand as an epitaph to the European civilization then destroying itself, Planck wrote: "Without the war I would never have known his value, and now that I know it, I must lose him."[89]

In 1917 Planck's daughter Grete, who had married a professor in Heidelberg, Ferdinand Fehling, died suddenly, a week after childbirth. Her twin, Emma, came to care for the infant, and in January 1919 she married the widower. By the year's end Emma too was dead after bringing a new life into the world. Her distracted father placed her ashes in the Grunewald cemetery, next to her sister's. "The two beloved children, who could not get along without one another in life, are now together for ever."[90] The double tragedy almost destroyed Planck. He wrote Lorentz: "Now I mourn both my dearly loved children in bitter sorrow and feel robbed and impoverished. There have been times when I doubted the value of life itself."[91]

Einstein saw Planck just after Emma's death. "Planck's

89. Respectively, Planck to Wien, 19 Aug. 1918 (AHQP), to Max and Emma Lenz, 24 July 1916 (MPG), and to Roethe, 29 July 1916 (Göttingen Univ. Library). Cf. Planck to Lorentz, 24 July 1917 (LTZ/6), and to Emil Fischer, 2 July 1917 (Fischer/28, TBL), re Erwin; Fischer to T. W. Richards, 16 Nov. 1914, in Reingold and Reingold, eds., *Science in America* (1981), 229, on the unexpected willingness, "which far surpassed our highest expectations," of German youth to die.

90. Planck to Emma Lenz, 6 Jan. 1919 and 3 Jan. 1920 (MPG), and to Frieda Clarke (niece), 23 Dec. 1919 (APS).

91. Planck to Lorentz, 21 Dec. 1919 (LTZ/6), and to Frieda Clarke, 23 Dec. 1919 (APS).

misfortune wrings my heart," he wrote Born. "I could not hold back the tears when I saw him. . . . He was wonderfully courageous and erect, but you could see the grief eating away at him." Planck found consolation, he wrote a niece, in a thought that became a dominant note in his lectures during the Nazi period: we have no right to all the good things that life brings us; unhappiness is the natural state of man. But not the inevitable state. "There are still many precious things on the earth and many high callings, and the value of life in the last analysis is determined by the way it is lived. So one comes back again and again to his duty to work on and to show those dearest to him as much love as he himself would like to experience."[92] He eventually found solace for the loss of his daughters in their children, whom he helped to raise.[93]

While these domestic and public tragedies took place, Planck's colleagues were awarding him the highest distinctions for his scientific work. In 1918, to mark his sixtieth birthday, the Physical Society held a session and published a pamphlet in his honor. Einstein organized the proceedings, as president of the Society for 1917–18. To Sommerfeld, whom he enlisted to write the major review of Planck's work, he wrote: "I'll be happy tonight if the gods grant me the gift to speak profoundly, because I am very fond of Planck, and he will certainly be very pleased when he sees how much we all care for him and how highly we value his life's work."[94] And this was not a sentiment he reserved for

92. Einstein to Born, 9 Dec. 1919, in Einstein and Born, *Briefwechsel* (1969), 39; Planck to Emma Lenz, 3 Jan. 1920 (MPG), and to Runge, 19 Dec. 1919 (RP).

93. HH, 17. The granddaughters were named after their mothers; Emma studied music, Grete became a physician. Planck to Frieda Clarke, 12 May and 23 Oct. 1939 (APS).

94. Einstein to Sommerfeld, ca. 1 Mar. 1918, in Einstein and Sommerfeld, *Briefwechsel* (1968), 48; Banesh Hoffmann, *Albert Einstein* (1972), 222.

celebrations. Einstein ended a list of the attractions Born would find if he returned to Berlin: "But chiefly this: to be near Planck is a joy."[95]

The advertisement of Planck's importance to physics probably helped to convince the Nobel prize committee to rectify what had become an evident injustice. As the committee itself admitted in its report of 1919, Planck had received more nominations, and from physicists representing a wider variety of specialties, than any other candidate. The leading continental theorists, Lorentz, Einstein, Born, Wien, and Sommerfeld, insisted upon him. Sommerfeld presented the committee with the account he had written for Planck's sixtieth birthday and emphasized that physics had become the physics of quanta.

Faced with the facts that the quantum theory had proved fruitful in many disparate branches of physics and that, as Laue observed, no prize could be given for any other work in quantum physics before Planck was honored, the committee waived the objection that Planck's initiative had not yet resulted in a coherent theory. In 1919 it put him forward for the unawarded prize for 1918. For the prize of 1919, for which Planck proposed Einstein for the general theory of relativity, the committee chose Johannes Stark, who in 1913 had shown by experiment that an electric field can alter the behavior of radiating atoms.[96] The Swedish Academy of Sciences accepted both recommendations and so simultaneously strengthened the hands of the spokesmen for two dia-

95. Einstein to Hedwig Born, 8 Feb. 1918, in Einstein and Born, *Briefwechsel* (1969), 23. Born shared the estimate: "You can certainly be of a different opinion from Planck's, but you can only doubt his upright, honorable character if you have none yourself." Born to Einstein, 28 Oct. 1920, ibid., 71, and 7 Apr. 1923, ibid., 110.

96. Nagel, in C. G. Bernhard et al., eds., *Science* (1982), 370–74; Sommerfeld to Nobel prize committee, 3 Jan. 1918, in "Förslag till utdeling av Nobelpriset i fysik," 1918, 211 (Nobel Archives, Kungl. Vetenskapsakademien, Stockholm).

metrically opposed factions of German physicists. Stark disliked theory and those who made it, as well as liberals and Jews and non-Germans in general.

It was an ill-assorted group that travelled together to Stockholm in the summer of 1920: Stark, anti-Semitic and paranoid, afraid that Planck might overshadow or ignore him; Planck, moderate and upright, collecting an honor universally recognized as his due; and Fritz Haber, winner of the chemistry prize for 1918, a liberal Jew and German patriot, almost universally condemned as the technologist of gas warfare. They were to play the parts of aggressor, peacemaker, and victim in the drama that destroyed German physical science. But for the moment they were a united band of heroes on the last battlefield on which Germany had a chance for glory. As the German ambassador to Sweden wrote to his ministry: "The whole business of the Nobel awards and festivities had the air of a homage to German science. . . . [It] will certainly contribute decisively to the advancement of cultural relations between our two countries, a point of particularly great importance in the current very unfavorable political situation."[97]

97. Planck to Stark, 28 Nov. 1919 and 17 Apr. 1920 (Stark Papers, SPK); German ambassador, Stockholm, to Auswärtiges Amt, Berlin, 8 June 1920 (Reichsministerium des Innern, Nr 8970, Bl. 120, Zentrales Staatsarchiv, Potsdam).

| 3 |

DOCTOR OF SCIENCE

Planck urged his fellow secretaries to keep up the regular meetings of the Academy during the time of troubles following the Armistice. The consequent exposure of academicians to the danger of revolution and general strikes counted for little, he declared, against the sacred duty to preserve "the place of German science in the world." Planck practiced his preaching. At the depth of his depression over Emma's death, he strove to cool a heated dispute within the Physical Society over what he called "a typical example of German eccentricity," the choice of a title for its new journal. Thus his colleagues wasted their energies squabbling among themselves while the great work of rebuilding lay waiting all around. Planck set them an example: whenever the frequent transit strikes in Berlin would otherwise have curtailed his work, he walked two hours each way from his suburban home to the city center to attend to reconstruction.[1]

Planck's unusual fortitude is thrown into relief by contrast with the bitterness of Wien, who wrote in his autobiography

1. Planck, *Sb* (1918), 993 (14 Nov.); Planck to Sommerfeld, 15 and 20 Dec. 1919 (AHQP/84); Laue, *Nwn*, *35* (1948), 5. Cf. Schroeder-Gudehus, *Minerva*, *10* (1972), 537–70.

that "a German of my generation, who has witnessed the decline and fall of his country, cannot be happy," and with the deep pessimism of Wien's reactionary friends, to whom Wien had occasionally to dispense a drop of cheer from his own small stock.[2] Planck held out hope to his colleagues. To Eilhard Wiedemann: "The main thing is not to lose courage and the hope that better days will come." To Carl Runge: "At the moment the outlook for our German science is very bleak. But I cling strongly to the hope that it will again reach the top (even viewed from outside), if only we can get through the next difficult years decently." To all, through the *Berliner Tageblatt*: "As long as German science can continue in the old way, it is unthinkable that Germany will be driven from the ranks of civilized nations."[3] Planck's personal qualities—his high German culture, his noble bearing, his recent canonization in Stockholm—and his tireless efforts to restore intellectual life in Germany brought him increasingly into the public eye. As other leaders of his generation of physical scientists left the scene—Emil Fischer died in 1919, Emil Warburg retired in 1922, and Adolf von Harnack died in 1930—Planck collected much of their authority and emerged as the chief spokesman for German science.

Of the Body

In 1919, at its annual commemoration of Leibniz, Planck exhorted the Berlin Academy to defend its treasure against a menace greater even than the ill will of the Entente powers. Its science, the still-proud legacy of the Empire and the means to the restoration of the state, faced ruin in the gen-

2. Wien, *Aus dem Leben* (1930), 50, 67, 69.
3. Planck to Wiedemann, 8 June 1919 (acc. Darmst. 1919.143, SPK), and to Runge, 2 Aug. 1919 (RP); Planck, "Internationale Arbeitsgemeinschaft," *Berliner Tageblatt*, 25 Dec. 1919, 4. Beiblatt.

eral declining financial situation.[4] It takes money to do science. The great inflation, beginning in 1919 and culminating in 1923, destroyed endowments of scientific institutions and made it impossible to purchase equipment or even journals from abroad and expensive to buy at home. The price of paper and labor drove up the cost of domestic books to the point that, as the president of the Naturforscherversammlung for 1920 joked painfully, they should be chained to desks as in the libraries of old. At the height of the inflation, the value of the mark changed so rapidly that in the course of a train ride away from Berlin the purchasing power of the sum given Planck for travel expenses on the Academy's business could sink beneath what he needed for a hotel. He would then, aged sixty-five, sit up all night in a station waiting room.[5]

In such circumstances self-help was the only help. Planck, Haber, Nernst, and other academicians constituted a commission to supervise a new Reichszentrale für naturwissenschaftliche Berichterstattung (National Center for Scientific Reporting), which was charged with securing at least one copy of every scientific book and journal not received in Germany during the war. As a beginning, they started looking for the thirteen thousand journals noted in the *World list of scientific periodicals* but not available anywhere within the new borders of the world's former leader in scientific bibliography.[6] Planck considered the activities of the National Center to be essential to the fight for Germany's place in the sun, and he encouraged the diffusion of reports of the foreign literature it amassed. "Especially we must support our review journals and reference works with all avail-

4. Planck, *Sb* (1919), 548. Harnack put the same point in similar words in 1920; Schlicker, *NTM*, *12:1* (1975), 45–46.

5. Friedrich von Müller, in GDNA, *Verhandlungen* (1920), 22; Laue, *Nwn*, *35* (1948), 5.

6. Richter, *Forschungsförderung* (1972), 26.

able resources so that they can take the competition from similar publications now cropping up everywhere in enemy countries on a German model."[7] Something more than being proud and keeping current was at stake. The winners of the competition, the authors of the handbooks of science, guide the writing of its history.

Science does not live by the word alone. Leading members of the Academy—Haber, Planck, Harnack—and the former Prussian Kultusminister Friedrich Schmidt-Ott planned a new organization, the Notgemeinschaft der deutschen Wissenschaft (Emergency Association of German Science), which would unify regional, disciplinary, and political factions in the common purpose of raising money for all the needs of German science. The conception linked the universalistic notion of science with the prevalent German organizational form of the cartel: the sponsors and members of the Association included the union of German academies, the universities and Technische Hochschulen, the Kaiser-Wilhelm-Gesellschaft, and the Gesellschaft deutscher Naturforscher und Ärzte (Society of German Scientists and Physicians). Planck, as presiding secretary of the Academy, stood briefly at the pinnacle of this pyramid and, on behalf of its constituents, invited Schmidt-Ott to lead it.[8]

The new organization began its official existence on 30 October 1920. It succeeded in winning aid from the central government, which had previously left the general support

7. Planck to Runge, 2 Aug. 1919 (RP); cf. Künzel, AW, ITGOW, *Kolloquien*, 35 (1984), 78.

8. AH, 61, quoting Planck to Schmidt-Ott, 19 Apr. 1920; LS, 2, 126–27; Forman, *Environment* (1967), 290–93, drawing on NDW, *Berichte*, 1 (1922), 13–16, 54–55. Over thirty years later Schmidt-Ott still considered it a wonder that all the governments, scientific organizations, and industrial firms approached agreed to adhere to the NDW without reservation—a proof, he said, of "how much the cultural heritage of our people was valued and its destruction feared, even in the changed political circumstances." *Erlebtes* (1952), 179.

of science to the individual states, from German industry, and from abroad. The Reich's contribution increased steadily to about 8 million marks in 1927 and 1928, of which 900,000 went for physics. The most extensive external help came from the Rockefeller Foundation, which by 1933 had distributed a half-million dollars through the Notgemeinschaft, about 5 percent of the Reich's total contribution and more than the contribution of German industry. The most important help for physics came from the Japanese industrialist Hajime Hoshi, whose donations, exhausted by 1925, supported investigations of the atom; and from an annual subvention of 12,500 dollars from the General Electric Company of the United States.[9]

Planck advanced the program of the Notgemeinschaft in his usual way, by attention to detail and by membership on key committees.[10] He sat on the Hauptausschuss, or executive board, on the committee to spend the Hoshi fund, and also on the Elektrophysik-Ausschuss (Electrophysics Committee), which administered General Electric's money together with 2,500 dollars donated annually and in equal measure by Siemens and Halske and the Allgemeine Elektrizitäts-Gesellschaft. Meanwhile the Notgemeinschaft's main committee on general physics came safely under the direction of Laue after the profession defeated a power play by Stark. (Stark had tried to have the steering committee of his breakaway rival to the Physical Society, the Forschungsgemeinschaft der deutschen Hochschullehrer [Research Association of German Professors], appointed as the physics committee of the Notgemeinschaft; instead Schmidt-Ott coopted Stark, who did not survive the first general election in which members of the Notgemeinschaft's

9. NDW, *Berichte*, *1* (1921–22), 32, *3* (1923–24), 33–36, 57–62, *4* (1924–25), 29, *12* (1932–33), 10; Richter, *Forschungsförderung* (1972), 35; cf. Forman, *Environment* (1967), 297, 301–4.

10. Cf. *PPA*, 75–78 (1924).

constituent societies chose its officers.) Under Planck's chairmanship, and with members Laue, Haber, Nernst, Max Wien, and, added later at Planck's request, James Franck and Friedrich Paschen, the Electrophysics Committee supported many projects in atomic and quantum physics. All the leading practitioners, including Born, Einstein, and Sommerfeld, received grants.[11]

The frequent awards to leading theorists and their many students may have made colorable Stark's charge that the Notgemeinschaft's physics committees followed "the mathematically oriented Jewish group, whose center is Sommerfeld [who was decidedly non-Jewish]." Planck's way of distributing the "very tidy sum" of fifteen thousand dollars a year from the Electrophysics Committee was not philo-Semitic, but pro-theorist and highly elitist. Tell me what you need, he wrote Sommerfeld, and I will see that you get it all. "To be sure, the money is designated 'especially for experimental research,' but your project can be presented as the working out of experimental research. The main thing of course is your name."[12]

In favoring atomic theory, Planck applied a policy at which he had hinted in a talk to the Kaiser-Wilhelm-Gesellschaft just after the war: since the quantum problem occupied the research front under attack by the physicists of all countries, there lay Germany's best chance to show that it still cultivated science with its old spirit and effectiveness.

11. Forman, *Environment* (1967), 313–19; Richter, *Forschungsförderung* (1972), 36–37, and Richter, *Sudhoffs Archiv*, *57* (1973), 196–200; Haber and Hahn, *Nwn*, *11* (1923), 31–32, 210–11; NDW, *Berichte*, *3* (1923–24), 35–36, 47–54.

12. Stark to Schmidt-Ott, 15 Dec. 1920, in Richter, *Forschungsförderung* (1972), 14; Forman, *Minerva*, *12* (1974), 62–63, quoting Schmidt-Ott, *Erlebtes* (1952), 180; Einstein to Born, 30 Apr. 1922, in Einstein and Born, *Briefwechsel* (1969), 103; Planck to Sommerfeld, 8 July 1923 (AHQP/84).

Here is the policy, as Planck stated it in 1920 to Vilhelm Bjerknes, a Norwegian meteorologist then prominent for his creative parallel between trench warfare and what he called a weather "front": "Obviously we can do nothing better toward [improving international scientific relations] than to work hard at science and so try to raise the value of exchanges with us." Against complaints that in furthering this policy his committee favored novelties, Planck replied, as he did to Wilhelm Wien in 1922, that the breakdown of the old, beautiful theories demanded steadfast inquiry into trouble spots. Shirking this duty would hurt the nation as well as its science. "Should we shrink from investigating these matters or ignore them, we would stagnate or fall behind other nations."[13]

In 1925 theorists who had received support from Planck's committee fashioned quantum mechanics. Planck promptly seized upon its creation as a justification of his policy for keeping German physics in the world's eye: "Quantum mechanics is at the center of interest of the physicists of all countries. And just there, in the work of Heisenberg and Born, which the Electrophysics Committee has supported and which without this support would most probably have had to be done elsewhere, it is clear how useful the Committee has already been for the development of physics in Germany."[14] Planck retained his membership on the Notgemeinschaft's executive committee until 1932 and his chairmanship of the Electrophysics Committee throughout the Weimar period.

Planck also served the other main financier of German science, the Kaiser-Wilhelm-Gesellschaft. He had become a senator of the Society in 1916 on the nomination of the kai-

13. *PA*, *3*, 119–20 (1919); Planck to Bjerknes, 27 Mar. 1920 (Oslo Univ. Library), and to Wien, 13 June 1922 (AHQP); AH, 64.
14. NDW, *Berichte*, *5* (1925–26), 85.

ser, who had held the right of appointment to half the seats.[15] Planck had immediately revived the question of a physics institute; with Harnack's help, he garnered enough in pledges—fifty thousand marks annually from the Society and half that from the Koppel Foundation—to underwrite an organization on a novel plan. It would have Einstein as its director and money for equipment and research, but no building; it would buy instruments and lend them to researchers in universities; it would, according to Harnack, strengthen and integrate research while reducing costs, thus realizing the purposes and exceeding the dreams of the science administrator.[16] Planck served as both an overseer and a director (a member of the Kuratorium and the Direktorium) of the new institute, which had opened its immaterial doors late in 1917. He often took care of its business when its nominal director was unable to do so.[17]

The Society hoped eventually to have a proper building for its physicists, but its dire circumstances in the immediate postwar years precluded financing from its own funds. Instead, as its executive secretary, Friedrich Glum, observed in 1921, when Einstein's organization had neither a scientific worker nor a technician, the brickless institute might have to be the pattern rather than the exception for the future activities of the Society. In the event, it changed not the form, but the content, of its institutes, preserving the older ones for the collection and production of the "general-staff offi-

15. KWG, Senat, "Protokoll," 31 Mar. 1916, 2, and 12 Feb. 1917, 2, 4; KWG, "Bericht über die 3. Hauptversammlung" (Apr. 1916), 2; KWG, *Liste* (1917), 17, 31 (all in Fischer/1, TBL).

16. KWG, Senat, "Protokoll," 11 May 1917, 5, 6 July 1917, 6–7, and 16 Jan. 1918, 5 (Fischer/1, TBL); Harnack, *Die Institute* (1917), 10. Half of the Society's portion came from a gift from "Fabrikbesitzer Stock in Stolzenburg."

17. KWG, *Liste* (1917), 31 (TBL); Planck to Einstein, 29 Dec. 1917, 20 July 1919, 5 Apr. 1922 (AE).

cers of science" and adding new ones for the captains of industry.[18]

Stabilization and expansion were secured by bringing in the state and industrial capital in a measure not foreseen in the Society's constitution. The appeal to both interests was the same: Kaiser-Wilhelm institutes would dig out the science and consider the applications needed for the reconstruction of German industry. After long negotiations, Prussia and the Reich agreed to split the shortfall in the Society's operating expenses, which amounted in 1920 to three times its income. In return, government acquired the former kaiser's right to choose half the Senate, now grown to thirty members. Planck played a part in the redrafting of the Society's statutes, in the preservation of its name, and in its general administration.[19] As for big industry, it paid its way by underwriting the creation of institutes dedicated to research of immediate utility in the manufacture and improvement of metals, textiles, and leather goods. Chemical interests assumed the running costs of the Society's chemical institute.[20]

Big industry came to have a preponderant influence in the Kaiser-Wilhelm-Gesellschaft, carrying into the peace the emphasis on research with quick payoff that had inspired the Society's rapid growth during the war, when Haber's chemical institute, to take an example, increased from 5 to 1,500 workers. Then, as Harnack said, industrialists came to un-

18. Glum, *Nwn, 9* (1921), 300, and Glum, in Harnack, ed., *Handbuch* (1928), 30; Einstein to Glum, 2 Mar. 1921 (MPG/A39).

19. KWG, Senat, "Protokoll," 3 June 1919, 15–16 (Fischer/1, TBL).

20. Harnack to Reichswirtschaftsamt, 28 Dec. 1918; KWG, "Protokoll," 11 May 1920; [Harnack] to the Preussischer Finanzminister, 20 May 1920; KWG, Hauptversammlung, "Bericht," 18 May 1921, all in MPG, *50 Jahre* (1961), 163–65, 169–70, 171–72, 176–78. Cf. Glum, in Harnack, *Die Institute* (1917), 17–18, 27–28, and Schlenck, in Abb, *Aus fünfzig Jahren* (1930), 295–98.

derstand that "scientific research is and will remain one of the first presuppositions of our military and material power."[21] By 1926 representatives of Krupp, big steel, and I. G. Farben held five places on the seven-member executive committee (Verwaltungsausschuss).[22] When economic conditions stabilized, about half the Society's operating expenses came from industrial and private sources and the other half from Prussia and the Reich in almost equal measure. Some two dozen new institutes were built, mostly to serve, and serve up, applied science.

No room was found in the expanding budgets of the Society for the erection of a proper physics institute: from the standpoint of the practical man, Germany already possessed what it needed in the research divisions of the Reichsanstalt.[23] This was a position doubly injurious to the cause of pure physics. It undermined efforts within the Society to establish a physics institute, and it menaced the work at the Reichsanstalt furthest from practical application. The menace increasingly became reality under Emil Warburg's successors Nernst (1922–24) and Friedrich Paschen (1924–33), an excellent precision spectroscopist without the scientific or political clout to protect his institution. Planck and Wien lamented that the Reichsanstalt was becoming more and more like the standards bureaus of Britain and the United States, emphasizing applied work and losing its ascendency, "which lay in its importance for pure research."[24] As the Reichsanstalt's science declined, the need for a Kaiser-Wilhelm In-

21. Harnack in KWG, "Mitteilungen für die Mitglieder," no. 8 (1916), 1, and Haber, report of 1917 (Fischer/1, TBL); cf. Harnack, *Die Institute* (1917), 7.

22. Reishaus-Etzold, *JW*, 1973:1, 43, 45, 48, 51–59; cf. Schlicker, *NTM*, *12:1* (1975), 48.

23. Reishaus-Etzold, *JW*, 1973:1, 39; Harnack, *Erforschtes* (1923), 246, 249, and Planck et al. to Harnack, 5 Mar. 1929, in Kirsten and Treder, *1*, 155, re the Reichsanstalt.

24. Planck to Laue, 8 June 1924 (DM).

stitute for Physics grew, at least in the minds of the Berlin physicists.

Planck frequently advised Harnack, and occasionally the Society, to keep its higher purposes before it, to eschew immediate practical payoff in favor of pure science, German culture, and international cooperation. Not all these ideals stood high among the priorities of Krupp or I. G. Farben. Nonetheless, to them Planck had merits that outweighed the disadvantage of his outmoded conception of the Society's purposes: he knew his way around the top organizations of German science, he enjoyed an unblemished character, and he had an international scientific reputation that might intimidate ministers of state. In July 1930, shortly after Harnack's death, the Senate chose its quiet, modest member Planck as its president. It did so in opposition to the government's candidate, the orientalist and sometime Prussian Kultusminister Carl Heinrich Becker, who favored stricter state supervision of the Society. He opposed the freewheeling policy of Harnack and Schmidt-Ott, under which industry received covert subventions in the form of state support to technologically oriented Kaiser-Wilhelm institutes. Apparently the Society's vice president, Krupp von Bohlen, who understood that Planck's coolness toward applied science would not cause him to try to alter the status quo, headed the move that defeated Becker.[25]

Fritz Haber compared the president of the Kaiser-Wilhelm-Gesellschaft to the king of England, who might overcome the weakness of his constitutional authority by the force of his personality. Apart from important policy issues, the main authority of the Society resided in the executive committee and its secretary, Glum; Planck's task was to deal with ministers and deputies, with men of commerce, bank-

25. Glum, *Wissenschaft, Wirtschaft und Politik* (1964), 377–78; Reishaus-Etzold, *JW*, 1973:1, 41; Krupp von Bohlen to Planck, 22 July 1930, in MPG, *50 Jahre* (1961), 189; Schmidt-Ott, *Erlebtes* (1952), 135.

ing, and industry, with journalists, diplomats, and foreign dignitaries—a task lightened, according to Glum, by Frau Planck's love of society. In the event, Planck's seven years in office were very burdensome. He began striving to mitigate the consequences of the worldwide economic depression for the Kaiser-Wilhelm institutes and ended working to protect the Society from the policies of the Third Reich.

The presidency of the Society elevated Planck from spokesman for German physics to (as he was variously styled) "the voice of German scientific research" and "the Nestor of German physical science." He wrote for the newspapers, submitted to interviews, and spoke on the radio. This last, untypical performance was mediated by Hans Hartmann, a pastor who had left his pulpit, taken to journalism, and married one of Planck's nieces.[26] One element in these communications may be remarked. Even when running a research enterprise some two-thirds of which, according to an estimate of 1941, had to do directly with application, Planck maintained the fiction, or rather the pretense, that the Society's institutes did not work for the industries that supported them. Its members pursued science as they pleased. "Of us it can truly be said: the spirit blows where it lists."[27] In this Planck remained true, at the price of truth, to his conception of a clear distinction between pure and applied science. To the one belonged the discovery of electromagnetic waves and the splitting of the atom; to the other

26. Heisenberg, *Nwn*, 23 (1935), 321, and Bachmann and Trummert, *Münchener medizinische Wochenschrift*, 112 (23 Jan. 1970), 159 (quotes); Planck, *Ernte*, 13:7 (Apr. 1932), 31–33; HH, 80–109 (transcripts of broadcast interviews in 1932 and 1935); Planck to Hans Hartmann, 25 Nov. 1932, 23 Jan. 1933, 30 Apr. 1934 (MPG). It is said that Planck's name appeared more often as an author in the *Vossische Zeitung* or the *Dresdener Anzeiger* than in the *Annalen der Physik*; D. Hoffmann, *Zur Geschichte* (1982), 21.

27. Bader, *Forschung* (1941), 121; HH, 90 (broadcast of 1935).

belonged the invention of radio and the possible exploitation of atomic energy; to neither, or to both, belonged most of the work of Planck's institutes and, for that matter, the measurements at the Reichsanstalt that had guided his way to the quantum theory.[28]

Planck agreed only reluctantly to take on the presidency of the Kaiser-Wilhelm-Gesellschaft. He still carried much of the immense administrative burden he had shouldered during the 1920s. Besides discharging his important offices at the Notgemeinschaft, he continued to be the dominant secretary in the Academy, stayed active in the Physical Society, refereed papers for the *Annalen der Physik*, and helped establish the Deutsches Museum.[29] Although Laue had lightened his load a little at the University by taking over the directorship of the Institute for Theoretical Physics in 1921, Planck continued to teach and examine after his retirement in 1926 (or rather 1927, since the faculty's first choice as his successor, Sommerfeld, declined to move from Munich and its second choice, Erwin Schrödinger, could not take up his duties immediately).[30] Planck appears to have continued his cycle of four weekly lectures through the academic year 1929–30, and he served on administrative committees of the

28. Planck, *Die Woche* (1931), and *Ernte* (1932), 33; HH, 94 (1935); Schlicker, *JW*, 1975:2, 162–63.

29. Planck to Sommerfeld, 1 July 1923, in AH, 68, and Ficker, "Max Planck zum Gedächtnis," *PB, 4* (1948), 163 (activities at the AW); E. Gruneisen, in *PA, 3*, 406, Planck to Sommerfeld, 2 Feb. 1929 (AHQP/84), and to Wien, 21 May and 25 Aug. 1925 (DPG and *Annalen der Physik*), 9 and 14 July 1922, 20 Mar. 1927 (DM), all in AHQP; Planck to Frieda Clarke, 2 May 1925 (APS), and exchange with Oskar von Miller, Feb. 1924, in Fuchs, *Der Aufbau* (1957), 57–58; Planck to Walter Gerlach, 14 Nov. 1928 (AHQP).

30. Planck to Wien, 29 May and 26 Aug. 1921, 1 Oct. 1926 (AHQP); to Laue, 10 Nov. 1927 (DM); to Max Schirmer, 25 Nov. 1927 (MPG); Benz, *Sommerfeld* (1975), 157–59.

University at least as late as 1932.[31] Retirement did not cur-
tail his attendance at the physics colloquia at the University
or the Reichsanstalt. Nor did it reduce his lecturing: it
merely gave him more time to travel around Germany to address wider and farther audiences. Planck evidently was an
exact economist with his time. A rigid schedule during
term—writing and lecturing in the morning, lunch, rest,
piano, walk, correspondence—and an equally unrelenting
recreation—mountain climbing without stopping or talking
and Alpine accommodation without comfort or privacy—
made it all possible.[32]

International Relations

The Nestor of science faced a number of problems for which
neither money nor exercise offered a solution. Many of these
concerned the resumption of international relations in science, the recovery of what Planck called the "dream picture
. . . of trusting international community life."[33] The problem was vexed by the official policy of the International Research Council, set up in 1919 by the Entente in place of the
prewar International Association of Academies dominated
by the Academy of Berlin. The statutes of the Council excluded nationals of the former Central Powers from its
administration and from meetings and projects arranged by
its constituent scientific unions.[34]

In justification of this policy its supporters often referred

31. D. Hoffmann, AW, ITGOW, *Kolloquien, 35* (1984), 57; Planck
to unnamed correspondent, 19 July 1932 (Autog. I/286, SPK).

32. Axel von Harnack, "Max Planck zum Gedächtnis," *PB, 4* (1948),
170–71; Ficker, ibid., 163–64; HH, 25; AH, 73–74. On the joys of climbing, see Planck's obituary of Paul Drude, *PA, 3,* 317 (1906).

33. *PPA,* 71 (1923).

34. Schroeder-Gudehus, *Scientifiques* (1978), 107–28, and Schroeder-Gudehus, *Deutsche Wissenschaft* (1966), 90–119.

to the ill-famed Appeal of the Ninety-three Intellectuals, which caused a disgust, in the words of Clemenceau, "greater than human language can express."[35] And this was not merely the overstated view of the French. In Lorentz's travels he found that whenever his acquaintances expressed their bitterness against Germany, they always gave "a prominent place to the unhappy manifesto of the ninety-three." Arrhenius also heard from many colleagues that the manifesto if not repudiated would hinder the resumption of international scientific relations. He therefore proposed to Emil Fischer in 1919 that all the signers still living do as Planck had done, and recant. Fischer granted that the signing had been a mistake and that its basis, belief in the truth of the German government's affirmations about Belgium, had been shattered by the facts. He talked the matter over with Haber, Nernst, Planck, and Waldeyer, all of whom applauded Arrhenius's initiatives and admitted their error but declined to take the initiative toward détente, "since we are all shocked by the peace treaty, which is so cruel to Germany."[36]

Planck no doubt had a strong voice in determining the reply. At Haber's suggestion he sent Fischer a copy of his open letter to Lorentz. In an accompanying note he counseled against following his example. "As things are now, I believe it would be entirely ineffectual both from a personal-moral and from a practical point of view to revert again, with any sort of explanation, to the Appeal of the Ninety-three Intel-

35. Speech of 21 Dec. 1918, quoted in Schroeder-Gudehus, *Scientifiques* (1978), 130; cf. Grundemann, in Dresden, Technische Hochschule, *Wissenschaftliche Zeitschrift*, *14:3* (1965), 802.

36. Lorentz to Einstein, 26 July 1919 (AHQP/86); Arrhenius to Emil Fischer, 31 Mar. 1919, and Fischer to Arrhenius, 17 Apr. and 12 May 1919 (quote), in Fischer/8, 6, 28, TBL, respectively. Cf. Planck to Vilhelm Bjerknes, 17 Mar. 1920 (Oslo Univ. Library): the cartel of German academies is prepared to resume exchange of communications with academies in Entente countries but will not take the initiative.

lectuals." Here, as in much else outside science, Planck disagreed with Einstein, who had proposed establishing a nongovernmental international committee to collect and evaluate the facts about the German invasion of Belgium in order to make repudiation of the Appeal inescapable.[37] Planck's opinion prevailed. The seventy-five of the ninety-three alive in 1920 were invited to recant. Of the fifty-eight who responded, forty-two expressed regrets and qualifications and the opinion that their retraction would be regarded in the Entente countries as cowardly or self-serving. In this, no doubt, they were right.[38]

At first the exclusionary policy of the International Research Council worked well: three-fourths of the international scientific meetings not held in Germany or Austria between 1920 and 1924 excluded Germans and Austrians. But the interest of the former neutrals, the British, and the Americans in sustaining the bitterness of the French and Belgians soon waned; and in 1925 Lorentz submitted to the Council's general assembly a proposal by the academies of Denmark, Sweden, and the Netherlands to cancel the exclusionary clause. The proposal lost.[39] The next year members of the Amsterdam Academy of Sciences spoke with Planck and two of his fellow secretaries about conditions under which Germans would accept membership if the Council would offer it. Planck and his colleagues replied that the invitation should be "positive" and accompanied by a declaration that the original reasons for exclusion were "an-

37. Planck to Emil Fischer, 25 Apr. 1919 (Fischer/28, TBL); Lorentz to Einstein, 4 May and 26 July 1919 (AHQP/56).

38. Wehberg, *Wider den Aufruf* (1920), summarized in Schroeder-Gudehus, *Scientifiques* (1978), 252, and Schroeder-Gudehus, *Deutsche Wissenschaft* (1966), 200–202; Planck to unnamed colleague, 3 Dec. 1923 (LTZ/8).

39. Schroeder-Gudehus, *Scientifiques* (1978), 133, 140–49; Grundmann, in Dresden, Technische Hochschule, *Wissenschaftliche Zeitschrift, 14:3* (1965), 799.

nulled"; that Germany and Austria should each have an appropriate proportion of offices in the Council and its unions; and that the German language should be allowed at meetings.[40] Meanwhile, however, bigger players had entered the game.

The great powers decided and contrived to lessen tension. The Locarno Pact dates from October 1925, the admission of Germany to the League of Nations from 1926. The Entente governments joined the neutrals and British and American scientists in bringing pressure on the Council, which dropped the offensive paragraph from its statutes in June 1926 and invited Germany and Austria to join the common front against the forces of nature. It did not, however, meet more than half of the first of the requirements set out by Planck and his colleagues. The German government encouraged acceptance of the invitation, although not instantly; the German academics played coy, demanded apologies, and ultimately refused their cooperation. They did not fully adhere to the Council until after World War II.[41] The several causes of this intransigence echo through Planck's Weimar writings.

Perhaps fundamental for men like Planck was the belief that science stands above politics. Its internationalist ideal, to which Planck returned again and again,[42] was both an earnest and an example of its transcendence; and just here the

40. C. G. Simpson to R. T. Glazebrook, 29 Mar. 1926, and to Arthur Schuster, 20 Apr. 1926, reporting a conversation with Lorentz. Cock, Royal Society of London, *Notes and records*, 37 (1983), 267.

41. Schroeder-Gudehus, *Scientifiques* (1978), 262–63, 266–72, 298; Grundmann, in Dresden, Technische Hochschule, *Wissenschaftliche Zeitschrift, 14:3* (1965), 805–6. According to Planck to Lorentz, 2 Feb. 1927 (LTZ/8), the central government forwarded the invitation to the academies and universities with the comment that no immediate reply was required.

42. For example, GDNA (1922), in HH, 52; *Sb* (1928), xvii; Planck to Lorentz, 22 Sept. 1925 (LTZ/8); *PPA*, 130 (1930); cf. LS, 2, 202, 206.

Western powers acted politically, the Germans scientifically. Planck reminded the Academy in 1926 that, despite provocation from abroad (and, he might have added, intransigence at home), it had not expelled enemy nationals during the war; and that the Entente had created a political organization, the International Research Council, where before the war a purely scientific one, the International Association of Academies, had reigned from Berlin. Later, in his necrology of Lorentz, he blamed the victors for having introduced "an unnatural mixture of science and politics, which cannot be justified on any objective grounds."[43] German scientists, whose national honor had been hurt by international ostracism, would not now cooperate unless the statutes of the Council explicitly excluded "political points of view and considerations that have nothing to do with science."[44] And, doubtless, Planck did not regard this condition or the imposing of it as a political act.[45]

The refusal of German scientists to adhere to the Council was political, not only in bringing pressure to win concessions or apologies from the Entente but also in expressing distaste for the Weimar Republic. They felt themselves betrayed by a regime that could sign the humiliating Treaty of Versailles, consider lowering admission standards for the universities,[46] and encourage acceptance of the crumbs of-

43. *Sb* (1926), lxxxviii; *PA, 3*, 348 (1928). The last echoes Planck's Leibniz Day talk of 1915, *Sb*, 482, and *PPA*, 28: "the fundamentally unnatural mixture of scientific and political activity."

44. Planck, *Sb* (1926), lxxxviii. Cf. Planck to Lorentz, 22 Sept. 1925 (LTZ/8): Germans prefer an organization of scientific entities, like the old International Association of Academies, to one of countries, like the International Research Council, "in which the political element naturally plays a much greater part."

45. Cf. Forman, *Isis, 64* (1973), 169–71; LS, 2, 97; *PhW*, 219.

46. A great worry in 1919–20; cf. Müller, in GDNA, *Verhandlungen, 86* (1920), 20; Ringer, *Decline* (1969), 206–7.

fered by the Council. Here again Planck took a middle, or rather a right-middle, and influential course, determined partly by his official responsibilities and partly by his political opinions. These latter were still represented by the Deutsche Volkspartei (German People's Party), whose leader, Gustav Stresemann, served as Reichskanzler in 1923 and then as foreign minister until his death in 1929.[47] The Volkspartei advocated political equality for all citizens; measures to strengthen the family and German racial stock; the reunion of separated Germans with the Reich; political, economic, and legal equality between the sexes; religious instruction in the schools and separation of church and state; freedom of the press; the cooperation of labor and management, and the right of labor to organize; and, in foreign affairs, it favored rebuilding the fleet, revising the Treaty of Versailles, regaining colonies, in short, reestablishing "the previous importance [of Germany] in the world."[48]

These policies went about as far as Planck could go toward social and political novelties, and he could not suffer the most democratic of them, universal suffrage. He thought it as reasonable for the masses to have the vote as for the inexpert to help decide between the theories of Newton and Einstein. Science would crumble if carried out by manifesto or majority rule; the state did just that, in Planck's view, by allowing promiscuous voting. He thus explained to Laue the origin of the miseries of the Third Reich: "The fundamental evil consists, in my opinion, in the coming of the dominance of the masses. Indeed, I believe the general right to vote (for twenty-year-olds!) to be a fundamental error." But, he con-

47. Apelt, *Weimarer Verfassung* (1946), 195; Bergsträsser, *Parteien* (1955), 256, 260–64, 277–78; Fricke, ed., *Die bürgerlichen Parteien* (1968), *1*, 646–57.

48. Fricke, *Die bürgerlichen Parteien* (1968), *1*, 653–54; Felix Salomon, *Die deutschen Parteiprogramme* (1920), 97–111.

tinued, the wheel of history cannot be turned back; he saw no purpose in playing the reactionary in the repellent Weimar Republic.[49]

In any case, the responsibilities of Planck's offices in the most powerful state-supported scientific institutions (Berlin Academy, Notgemeinschaft, Kaiser-Wilhelm-Gesellschaft) were incompatible with opposition to the established order. In 1924 he publicly defended the government against an attempt to lay upon it blame for the financial plight of German science. Also, Planck's son, Erwin, who had an important place in the Reichswehrministerium (Ministry of Defense), perforce supported the regime, and Erwin and his father were very close. With the inauguration of the highly conservative General Paul von Hindenburg as president of the Republic in 1925, Planck became almost enthusiastic about the political climate. "I hope and trust [he wrote a niece] that things will go well under Hindenburg and that we shall see gradual progress in the reestablishment of peace and order."[50] Nonetheless, he could not bring himself easily into line with the regime's policy toward foreign overtures if he spied in them any trace of "political" motivation.

Just before the International Research Council lifted its boycott, Planck stated what he took to be the correct policy to Wien, anent the meeting of the British Association for the Advancement of Science of 1926, to which he had received no special invitation. "But I would not have any reservations at all if someone accepted such an invitation, provided that it applied not to the individual person, but to German sci-

49. Planck to Laue, 31 Aug. 1943 (DM); Kangro, *Dictionary of scientific biography*, *11*, 14, s.v. "Planck"; Planck, *Das Reich*, no. 32 (9 Aug. 1942), on the elitism of science.

50. Planck, *Sb* (1924), xvii; LS, *2*, 201; Schlicker, *JW*, 1975:2, 180; Planck to Emma Lenz, 25 May 1924 (MPG), to Frieda Clark, 2 May 1925 (APS), and to Staatssekretär Pünder, 6 June 1927 and 2 July 1929 (Bundesarchiv, Koblenz).

entists [in general]." An individual might properly attend a meeting open to all relevant experts, but not a closed conference from which otherwise eligible German scholars were excluded on political grounds.

Just after the Council opened itself to the Central Powers, Planck received two invitations that put his policy to the test, one to a congress at Como, where fascist Italy proposed to do honor to the centennial of the death of Alessandro Volta, the other to the Solvay conference of 1927, the first after the war to which any Germans other than Einstein were invited. To the Italians, who had exercised no evident test but eminence, Planck gave a provisional acceptance, conditional, he told Wien, on how the political winds were blowing over Lake Como before the congress. But the apparent openness of the invitation made its acceptance almost a duty: "German scientists would fall under the suspicion of harboring bad will if they held back from a really purely scientific affair without very strong reasons." Planck attended and was satisfied; in his judgment the congress, which heard several speakers praise the fascist state and ended in a party given by Mussolini, went off "entirely harmlessly from a political point of view."[51]

The Solvay affair presented a much more difficult problem. Planck multiplied scruples in his letters to Lorentz: he no longer stood in the front ranks of quantum theorists, he could not speak any language but German, etc.; but the sticking point—"something uncommonly oppressive to me"—was that the Solvay authorities appeared to have applied a political test in omitting Sommerfeld, who had contributed much more to physics in recent years than Planck had. Planck suspected that Sommerfeld had not been invited because the Institut Solvay of Brussels did not want the

51. Planck to Wien, 6 Mar. and 29 Nov. 1926, 21 Oct. 1927 (AHQP); Congresso internazionale dei fisici, *Atti* (1928), *1*, ii–ix, 7, 49.

company of a German who had advocated the annexation of Belgium. Here Planck diagnosed the cause but mistook the symptom: the politics lay not in omitting Sommerfeld, but in including Planck. The original plan provided for only three Germans, one of whom had to be Planck as a symbol of healing and continuity; for the remaining two places Heisenberg, Born, and Pauli, all prime contributors to the new quantum mechanics that constituted the subject of the meeting, had sufficiently strong claims to bar Sommerfeld without recourse to politics.[52]

Not until after Planck visited Holland in late May 1927 to receive the newly instituted Lorentz medal of the Amsterdam Academy of Sciences, an award decided by people concerned to renormalize international scientific work, did he decide to go to Brussels.[53] No doubt Lorentz, with whom he stayed, acquainted him with the facts of the case and helped him to see the symbolic importance of his presence. Lorentz worked by small, positive steps, not by ineffectual gestures, toward reestablishing international scientific relations.[54]

Planck's attitude toward the resumption of formal relations was probably more representative of German natural scientists than the hard line urged by the Verband deutscher Hochschullehrer (Union of German Professors) and by some technologists and industrialists. In 1923 Karl Kerkhof, a leading spokesman for the Union and chief of the National Center for Scientific Reporting,[55] requested that the Weimar government withdraw from the international scien-

52. Planck to Lorentz, 13 June 1926 (quote), 2 Feb., 6 June, 18 Dec. 1927 (LTZ/8); Lorentz to Einstein, 6 Apr. 1926 (AHQP/86).

53. Planck to F. A. F. C. Went, 30 Apr. 1928 (Museum Boerhaave, Leyden).

54. Lorentz's pace was too slow for the majority of the Berlin Academy, which refused to print appreciative remarks that Einstein made about him. Planck to Einstein, 4 Dec. 1925 (AE); cf. Wien to Planck, 1 May 1915 (AHQP).

55. Schlicker, *JW*, 1975:2, 169.

tific organizations in which the Versailles Treaty allowed German membership (one being the International Committee on Weights and Measures) and that it not send delegates to any future international scientific congresses. The government forwarded the request to eight scientific institutions: two replied favoring Kerkhof's proposal (the Prussian Aeronautical Observatory and the National Center for Earthquake Research), one abstained (the Kaiser-Wilhelm-Gesellschaft), and five opposed, including the Notgemeinschaft and the Reichsanstalt.[56]

The refusal of German science to join the International Research Council after being invited to do so did not reflect a change in the balance of opinion but the fact that German scientists could forgo adherence without injury to their disciplines.[57] As Schmidt-Ott had said in 1920, they would show the victor his error, that "the sacred flame, the great passion, the life-giving force of scientific research" had not diminished in Germany, that the world could not do without German science.[58] The success of German research in the early 1920s, which had produced quantum mechanics among other things before the Council opened its doors, confirmed that German scientists did not need the organizations of international science and could indulge their domestic politics without laming their work.

In this last opinion they were perfectly right. As Planck insisted, the most natural and useful form of international

56. Grundemann, in Dresden, Technische Hochschule, *Wissenschaftliche Zeitschrift*, *14:3* (1965), 804. On the course of the so-called "Counter Boycott," see Schroeder-Gudehus, *Scientifiques* (1978), 272–98, Schroeder-Gudehus, *Deutsche Wissenschaft* (1966), 213–65, and Schlicker, AW, ITGOW, *Kolloquien*, *24* (1981), 33–37.

57. Cf. Cock, Royal Society of London, *Notes and records*, *37* (1983), 269, 273–74.

58. Schmidt-Ott, in NDW, *Berichte*, *2* (1922–23), 4, and ibid., *4* (1924–25), 5: "There is no better answer to [a recent affirmation of exclusion] than scientific progress in Germany." Cf. HH, 52.

cooperation emerges from substantive, scientific initiative, not from the dictates of scientific bodies.[59] "That is the way I have always tried to act and how I hope to be able to continue." One of his theaters of action was the Deutsche Kommission für die geistige Zusammenarbeit (German Commission for Intellectual Cooperation), the national affiliate of the League of Nations' Commission internationale pour la coopération intellectuelle, where he promoted, among other things, collaboration among the world's major libraries.[60] Individual German scientists, responding to other substantive innovations, accepted invitations to international meetings, lecture engagements, or study trips abroad as "informal" occasions that did not breach solidarity. Even the constituent unions of the International Research Council, which had not always followed the exclusionary policy of their parent before 1926, enjoyed the unofficial participation of German savants.[61]

Another instance in which German scientists conducted a foreign policy not entirely in keeping with their government's occurred in 1925, on the occasion of the two hundredth anniversary of the Russian Academy of Sciences. The Soviet regime, interested in employing German resources and engineers to further its program of industrialization, invited German institutions to send delegates to the anniversary celebrations in September. Despite the Treaty of Rapallo of 1922, the Weimar government did not welcome

59. Planck, *Sb* (1926), lxxxvi, xc; Planck to Lorentz, 2 Feb. 1927 (LTZ/8); *PA, 3*, 348 (1928); Planck to F. A. F. C. Went, 4 Feb. 1928 (Museum Boerhaave, Leyden); Planck to Einstein, 23 Mar. 1933 (AE); Ehrenfest to Planck, 26 Mar. 1933 (ESC); Clark, *Einstein* (1971), 459–60.

60. Planck to Paul Langevin, in André Langevin, *Paul Langevin* (1971), 123–24; Planck to Karl Brandi, 2 Mar. 1931 (Göttingen Univ. Library).

61. Forman, *Isis, 64* (1973), 179–80; Schroeder-Gudehus, *Scientifiques* (1978), 235–39, 286–87.

a show of interest in the Soviet Union on the part of its scientists, especially in the fall of 1925, when Stresemann was deep in the negotiations that would eventuate in the Locarno Pact in October. As Weimar turned west, its scientists turned east.

At first the government made its support of the German delegations conditional on reconsideration by the Soviet regime of death sentences passed against two German nationals for acts of terrorism; the scientists did not like being pawns in this game, even if the objective was humanitarian, and many suspected, rightly apparently, that the "terrorists" were German agents.[62] Eventually the Foreign Office concluded that reluctant German participation would undermine national goals and, working through the Academy and the Union of German Professors, endorsed the visit to Moscow. A sizable German delegation attended under the leadership of Schmidt-Ott and Planck. The right-of-center parties to which Planck and Schmidt-Ott adhered protested against such flagrant fraternization with the Bolsheviks.[63]

Planck had favored acceptance of the Soviet invitation from the beginning. He figured prominently at the celebrations, which, to sound a nonpolitical note, opened with the singing of the Internationale and closed with the chorus from Beethoven's Ninth Symphony. A banquet for 1,500 people followed, at which Planck spoke, praising the products of academies as the best relaxants for a world of political and economic tensions; "There is no bond [he said, ignoring

62. Ludloff, in Jena, Univ., *Wissenschaftliche Zeitschrift*, 1956–57:6, 709–15; Schroeder-Gudehus, *Scientifiques* (1978), 294–95; Forman, *Isis*, *64* (1973), 167.

63. Grau, in Bielfeldt et al., eds., *Deutschland, Sowjetunion* (1966), 173, 175–77; Schmidt-Ott identified with the rightist opposition Deutsche Nationale Volkspartei. *Erlebtes* (1952), 166. As late as 21 June 1925, Planck did not know whether the Academy would send anyone (letter to Wien, AHQP).

the experiences of the war] that unites the different countries of the earth, with their widely divergent interests, so directly and so harmoniously as science." In a phrase he would never have used at the Academy, he enrolled scientists under the banner of workers: "We work to set out the treasure of pure knowledge, of truth, at the roots of ignorance and superstition."[64] In subsequent negotiations Planck recommended investment in pure rather than applied science; Schmidt-Ott agreed to certain exchanges and joint expeditions, one of which named a landmark in central Asia the "Notgemeinschaft-glacier"; and the Notgemeinschaft set up a committee for cooperation with Soviet science, on which the ubiquitous Planck sat.[65]

Planck's participation in the opening toward the U.S.S.R. illustrates the difficulty, or perhaps the impossibility, of acting under even the highest, supranational, "unpolitical" academic ideals without in fact making a political statement. He rationalized the matter thus to Lorentz: "Although of course the whole celebration was not without political overtones, still it had mainly a scientific character." He gave as an example the entirely political decision of the Soviet regime to support pure science as "an eminently 'useful' activity."[66] Here Planck's idealism obscured his wisdom. He had played a political part in the Soviet Union.

Similarly, closer to home, his championing of pure over applied science, especially within the Academy, where he took the hierarchy to be a matter of definition, had significant political implications. As he said in 1922, the Academy

64. Quoted by Vogel, in Bielfeldt et al., eds., *Deutschland, Sowjetunion* (1966), 468; Schmidt-Ott, *Erlebtes* (1952), 221–22; LS, *1*, 218.

65. Schmidt-Ott, *Erlebtes* (1952), 226; LS, *2*, 221; Pachaly, in Bielfeldt et al., eds., *Deutschland, Sowjetunion* (1966), 132–34. Cf. Schroeder-Gudehus, *Scientifiques* (1978), 247–48, summarizing NDW, *Berichte, 5* (1925–26), 127, 7 (1927–28), 173.

66. Planck to Lorentz, 22 Sept. 1925 (LTZ/8).

had the advantage over sister institutions like the universities of insulation from the affairs and pressures of daily life— whence its freedom to do its job "in an unprejudiced, tireless struggle for the steady increase of pure knowledge, the most precious prize of pure scientific work." Shortly before he made this statement, Planck had joined his fellow secretaries in rejecting a request from the Reichsverband deutscher Technik (National Union of German Technology) for inclusion of a technical class within the Academy on the ground that its members would constitute a foreign body, destructive of the Academy's bond of pure, disinterested scientific research. From time to time, he reminded the academicians that the enduring contributions of their founder, Leibniz, lay not in his practical but in his philosophical and mathematical work.[67]

In the temper of the time, rejection of the claims of applied science to parity with pure science expressed something more than academic snobbery. Stark's Fachgemeinschaft, the ineffectual rival to the Physical Society, cultivated close ties with industry. Stark himself resigned his professorship in Würzburg in 1922, ostensibly because the faculty rejected as too technological a student's thesis on the optical properties of porcelain and perhaps also because he hoped to have the presidency of the Reichsanstalt to which, to his disgust, Nernst was appointed. The income from his Nobel prize had made retirement possible; Stark left academic life to improve porcelain and make trouble.[68] In identifying himself with industry he made common cause with Philipp Lenard, likewise a Nobel prizewinner, excellent ex-

67. *PPA*, 48 (1921); LS, *2*, 114–15; Schlicker, *JW*, 1975:2, 171–72; *Sb* (1935), cii. The Academy did admit technical physicists who had also done "scientific" work, for example, Karl Willy Wagner and Johannes Stumpf, in 1926; cf. Planck, *Sb* (1926), xcv ff., and LS, *2*, 236.

68. Planck to Nernst, 23 June 1922 (DM), and to Wien, 31 Aug. 1921 (AHQP); ADB, 11–12.

The impact of press media in Germany is very much repeated in the U.S. today. This is a common phenomenon.

perimentalist, and frustrated, incompetent theorist. In 1924 both became public champions of Hitler, whom they likened to Galileo, Kepler, Newton, and Faraday.[69]

Anti-Antirelativity

Opposition between pure and applied physics played a part in "Deutsche Physik," a program and a slogan for a physics free from highly abstract and mathematical theory. The goal of increasing the prestige of experimental physics had some appeal for technologists still at home with the method of cut-and-try and also for anti-Semites who identified relativity and quantum theory as the decadent work of Jews. Planck had been drawn into this ugly business from its beginning. After British astronomers confirmed predictions of general relativity and made its author a universal hero, Einstein became a target for the right-wing press. By enlisting the support of a few prominent physicists, particularly Lenard, and others with technical training like Stark's student of porcelain Ludwig Glaser,[70] the attackers gave the impression that Einstein's science was as infected as his political opinions and religious origins.

The ringleader of the attack was Paul Weyland, a professional troublemaker and small-time criminal, supported, according to the conjecture of the German Foreign Office, by anti-Semitic industrialists like Henry Ford. Weyland announced a series of twenty public lectures to be given in the fall of 1920 under the auspices of his "Arbeitsgemeinschaft deutscher Naturforscher zur Erhaltung reiner Wissenschaft" (Study Group of German Scientists for the Preservation of Pure Science). Only one meeting took place, on 24 August 1920, despite Weyland's announcement that lecturers would

69. Richter, *Sudhoffs Archiv*, 57 (1973), 200; Pyenson, *Young Einstein* (1985), 232; ADB, 95.
70. ADB, 11–12.

earn ten or fifteen thousand marks. Auditors had the opportunity to buy swastikas and anti-Semitic literature and to hear Weyland slander Einstein for plagiarism, publicity seeking, and scientific dadaism. Einstein himself attended for entertainment, in the company of Nernst, but the viciousness of Weyland and the unfair account of relativity in the lecture by the physicist Ernst Gehrcke did not provide him much amusement.[71]

Laue, Nernst, and Rubens responded with a brief and dignified notice in a leading Berlin daily, testifying to the depth of Einstein's work and the loftiness of his character, generous, modest, and opposed to every form of self-advertisement. Einstein also answered in the newspapers, against the advice of Born and others who worried that he would subject himself to further attack and appear to lower himself to Weyland's level.[72] The most significant and instructive response to Weyland's performance came from two scientific corporations in the formulation of whose policies Planck had an important voice, the Academy and the Society of German Scientists and Physicians.

The Academy decided to remain silent despite pressure from some of its members and from the Kultusminister. The interest of the minister, who wrote Einstein "with feelings of sorrow and bitterness" over the slanderous attacks in the rightist press, had perhaps been aroused by the Foreign Office, whose representative in London had reported the indignation in Britain over the attacks and the rumor circulating there that Einstein might leave Berlin. And that, the

71. Grundemann, in Dresden, Technische Hochschule, *Wissenschaftliche Zeitschrift*, *16:5* (1967), 1624–25; Laue to Sommerfeld, 25 Aug. 1920, in Einstein and Sommerfeld, *Briefwechsel* (1968), 65.

72. Laue, Nernst, and Rubens, *Tägliche Rundschau*, 26 Aug. 1920, in Kirsten and Treder, *1*, 208–9; Hedwig Born to Einstein, 8 Sept. 1920, and Einstein to the Borns, 9 Sept. 1920, in Einstein and Born, *Briefwechsel* (1969), 58–59.

representative judged, would be a catastrophe, not for German science, for which he had no official concern, but for German foreign relations. "Professor Einstein is just now a cultural factor of the first importance for Germany, since Einstein's name is known far and wide. We should not drive away such a man, whom we can use in effective cultural propaganda." And, indeed, Einstein, who had opposed the war and the manifesto of the ninety-three, was one of the very few German scientists invited to lecture in the Entente countries in the first postwar years. He himself recognized his value in this respect when, having refused an invitation to visit Paris out of solidarity with his Berlin colleagues, he was persuaded by the foreign minister, Walther Rathenau, to go in the interests of Franco-German relations.[73]

The scent of politics caused the Berlin academicians to avert their noses from the plight of their illustrious colleague. As Gustav Roethe, then the sitting secretary, wrote Planck, public silence is best in "an essentially political matter." Also, although it would not have occurred to Roethe to put it this way, the Academy had become too politicized to act unanimously in support of Einstein, "since the tasteless glorification of the new Archimedes and Newton was distasteful to so many." Planck concurred in Roethe's decision and added that Weyland and his supporters were too small a target for the big guns of the Academy, especially since the main point had been gained: Einstein intended to stay in Berlin.[74]

A statement in favor of Einstein of the kind issued by

73. K. Haensch to Albert Einstein, 6 Sept. 1920; F. Sthamer to the Foreign Office, 2 Sept. 1920; Einstein to the Berlin Academy, 13 Mar. 1920; in Kirsten and Treder, *1*, 204, 206–7, 210, respectively. Cf. ibid., 225–40; LS, *2*, 99–102; Planck to Sommerfeld, 1 July 1923 (AHQP/84): "I am convinced that by that [a trip to the United States] you have rendered a good service to German science and also to German politics."

74. Roethe to Planck, 10 Sept. 1920, and Planck to Roethe, 14 Sept. 1920, in Kirsten and Treder, *1*, 205–6.

Laue, Nernst, and Rubens would be "political" if made by the Academy, but indirect support by a discussion of relativity in a scientific setting was perfectly acceptable and even desirable. This is the course that Planck urged as a member of the executive committee of the Society of German Scientists and Physicians. In accordance with this advice and a proposal from Sommerfeld, who had once described relativity theory as a glory of Teutonic philosophy, the Society arranged a joint session of its mathematics and physics sections at its meeting in Bad Nauheim in the week of 19–25 September 1920. The point of the collective session was, as the Society's president, instructed by Sommerfeld, put it in his opening address: "Scientific questions of such difficulty and of such great importance as the theory of relativity cannot be voted on in popular meetings with demagogic slogans and in the political press with venomous personal attacks; in contrast, within the tight circle of true experts they receive an objective evaluation that does justice to the importance of their ingenious creator."[75]

Planck, who condemned (but not publicly) Weyland's actions as "scarcely believable filth,"[76] presided over the session at Bad Nauheim. It culminated in an exchange between Einstein and Lenard. Planck preserved the proprieties. Lenard's objections came down to a matter of taste; for him physics had to be intuitive, for Einstein abstract.[77] His performance pleased the Kultusminister and kept alive Sommerfeld's belief that Lenard had not gone over to the Weyland group; but this was to ignore Lenard's increasing anti-

. 75. Sommerfeld to Einstein, 3 Sept. 1920, in Einstein and Sommerfeld, *Briefwechsel* (1968), 65, 68; Sommerfeld, *Deutsche Revue*, *43:3* (1918), 132; Müller, in GDNA, *Verhandlungen* (1920), 17. Cf. Planck to Einstein, 22 Oct. 1921 (AE).

76. Planck to Einstein, 5 Sept. 1920 (AE).

77. *Physikalische Zeitschrift*, *21* (1920), 666–67; Weyl, in Deutsche Mathematiker-Vereinigung, *Jahresbericht*, *31* (1922), 61–62.

Semitism, his unbalanced judgment, and his malevolent character, "put together [as Einstein had diagnosed it as early as 1910] entirely of bile and intrigue."[78]

Planck, too, had known for a long time that Lenard was unstable.[79] Characteristically, however, Planck could not bring himself to oppose the academic advancement of either Stark or Lenard on political grounds. He had the opportunity in 1924, when Gehrcke put forward both as possible successors to the still unfilled professorship of experimental physics vacated by Rubens's death in 1922. The selection committee, which included Planck, Laue, and Haber, had no trouble rejecting the proposal. Its reason: "Because of their passionate and not always objective opposition to the new theoretical physics, these important scientists would endanger the fruitful collaboration of the Berlin physicists." Planck's reason, as he explained it to Wien: "He [Lenard] confuses subjective intuitions with objective facts, believes that he grasps material he does not understand, and does not recognize the limits of his importance. That is very dangerous for an academic teacher." Not a word about passion or politics; only the objective objection that Lenard would be a poor teacher.[80] Planck regretted that Lenard's and also Stark's professional incapacity ruled them out, since the faculty's choices, Wilhelm Wien, Max Wien, and James Franck (whom Planck considered the most important of the younger experimentalists), all declined.[81]

78. Haensch to the Foreign Office, 29 Sept. 1920, in Kirsten and Treder, *1*, 208; Sommerfeld to Einstein, 3 and 11 Sept. 1920, in Einstein and Sommerfeld, *Briefwechsel* (1968), 65, 71; LS, *1*, 116; Einstein to J. J. Laub, 11 Nov. 1910 (AE), as quoted in Pyenson, *Young Einstein* (1985), 232.

79. Planck to Wien, 9 Feb. 1911 (AHQP).

80. Haberditzl, in Harig and Mette, eds., *Naturwissenschaft-Tradition-Fortschritt* (1963), 321; LS, *3*, 172; Planck to Wien, 19 June 1926 (AHQP).

81. Planck to Wien, 20 Nov. 1922 (Autog. I/1282, SPK); D. Hoffmann, AW, ITGOW, *Kolloquien*, *35* (1984), 14.

The situation was resolved by an elaborate plan of exchanges, whereby Nernst would resign the presidency of the Reichsanstalt to take Rubens's chair, Planck would take the presidency of the Reichsanstalt, and Planck's chair would go to someone like Born. Planck decided not to seek the presidency, however, and the government appointed Paschen. Some more knots in this tangled affair: Stark had protested Nernst's call to the Reichsanstalt in 1922, both to the Ministry of the Interior and to Planck as a senior member of the Reichsanstalt's advisory committee (Kuratorium); Planck supported Nernst; Stark could not understand Planck's refusal to help him to a post at Berlin; Planck regarded it as a pity that Stark's disabilities tied his hands; and Stark eventually succeeded Paschen at the Reichsanstalt.[82]

That matters went so well at Bad Nauheim was largely the work of Planck, who prevented hecklers from interrupting Einstein and refused to allow a speaker who referred to articles in newspapers to continue. As Hermann Weyl, who was present, summed it up: "Planck discharged his office of chairman with great ability, strength, and impartiality; it was owing largely to him that the 'Nauheim discussion of relativity,' in which opposing conceptions of the epistemological foundations of science confronted each other, took place in a dignified way."[83]

On the day after the meeting, Weyland printed an article that called for a renewed attack on relativity and its defenders, on "this rat's nest of scientific corruption." From this it can be gathered that the encounter at Bad Nauheim was a

82. Haber to Staatssekretär H. Schulz, June 1924, re Planck's move to the PTR, in Kirsten and Treder, *1*, 162; Planck to E. Lescher, 17 Sept. 1924 (APS, quote); Stark to Planck, 10 Sept. 192? (SPK); Stark to Reichsministerium des Innern, 6 June 1922, and Planck to W. Nernst, 23 June 1922 (DM).

83. Meitner, *Nwn*, *45* (1958), 407; ADB, 90, quoting recollections of Felix Ehrenhaft; Weyl, in Deutsche Mathematiker-Vereinigung, *Jahresbericht*, *31* (1922), 51. Cf. Hermann, *New physics* (1979), 59–60; Schmidt-Ott, *Erlebtes* (1952), 181.

meeting by and for professional physicists, and that it had little effect on the sort of audience Weyland addressed. Building on the shaky agreement among physicists reached at and after Bad Nauheim "with great effort and many compromises," Planck, as president of the Society of German Scientists and Physicians for 1922, proposed that Einstein be asked to address a general session at the annual meeting. As he wrote Wien, he hoped that the lecture, which was to be a main event in the Society's centennial celebration, would replace "the silly advertisements that puff relativity theory" with "a purely objective point of view."[84]

Einstein was invited and accepted. He then withdrew, frightened by threats against his life, which, in the days after the murder of his friend and coreligionist Rathenau, he had to take seriously.[85] This foul intervention of the outside world in the doings of pure science disgusted Planck. "We are lucky to have come so far that a band of murderers, who go about their business in the dark, dictate the scientific program of a purely scientific body." He enlisted Laue to take Einstein's place, which, he admitted, had an "objective" (by which he meant "political") advantage. "Looked at entirely objectively, this change has perhaps the advantage that people who still believe that the principle of relativity is basically Jewish propaganda for Einstein . . . will learn [the truth]."[86]

The advantage did not disarm the ultras. Stark distrib-

84. Grundemann, in Dresden, Technische Hochschule, *Wissenschaftliche Zeitschrift*, 16:5 (1967), 1624; Planck to Wien, 13 June 1922 (AHQP). Cf. Sommerfeld to Einstein, 7 Oct. 1920, in Einstein and Sommerfeld, *Briefwechsel* (1968), 72.

85. Einstein to Planck, 6 July and 12 Aug. 1922, and Planck to Einstein, 8 July 1922 (AE). Rathenau was killed on 24 June 1922.

86. Planck to Wien, 9 July 1922 (AHQP), and, in almost the same words, to Laue, same date (DM); cf. Hermann, *PB*, 29 (1973), 483–87. Einstein had found it prudent to withdraw from an earlier scheduled appearance, at the Naturforscherversammlung of 1919, for similar reasons; Planck to Laue, 8 July 1919 (DM).

uted a poisonous pamphlet entitled *Die gegenwärtige Krisis in der deutschen Physik* (*The contemporary crisis in German physics*). It attacked theory, Jewish theory, and Einstein, and it advised physicists to go to industry, "like Johannes Stark."[87] Einstein went to Japan. His absence did not appease the anti-Semites, nor did the news that he had been awarded retroactively, in 1922, the Nobel prize in physics for 1921. Lenard complained to the Swedish Academy of Sciences and its committee on the physics prize that they "had not been able to bring to bear a sufficiently clear Germanic [!] spirit to avoid perpetrating such a fraud."[88] In 1923, with Einstein in refuge in Holland, Planck tried to get to the bottom of the plot against him. Neither Jewish leaders nor public officials (Planck put his inquiries even to health officers concerned with the insane) could tell him anything new. Planck decided that Einstein had been elevated as a lightning rod for discharging the enmity of people aroused by publicity about Jewish black marketeers "and because of reports about the attempted coup in Munich, which was as criminal as it was shortsighted."[89]

Planck was indignant at the anonymous threats. Stronger still than his indignation, however, was his worry that the rabble would finally drive away the ornament for which the entire world envied Germany. He wrote to Paul Ehrenfest, with whom Einstein was staying, that the Academy wished to put no pressure on Einstein, that he could stay away or do what he pleased, provided only that he maintain his official domicile in Berlin and give at least one scientific lecture there a year.[90] Although Einstein returned to his friends, his

87. Sommerfeld to H. O. Wieland, 27 Oct. 1928 (DM).

88. Letter of 14 Jan. 1923, copy in DM.

89. Planck to Einstein, 10 and 18 Nov. 1923 (AE). Cf. Planck to Einstein, 20 July 1919 (AE).

90. Planck to Ehrenfest, 30 Nov. 1923 (ESC); Planck to Lorentz, 5 Dec. 1923 (LTZ/8); Forman, *Isis*, 64 (1973), 166–67.

sinecures, and his sailboat, the outcome Planck feared was only postponed until the shortsighted criminal of the Munich coup gained control of the country. There was to be no retreat of the kind that Hedi Born urged on Einstein, "into the tranquil temple of science."[91]

Of the Soul

During the 1920s Planck reaffirmed and reworked his philosophy in response to general problems of Weimar society and to rapid progress in quantum theory. It took physicists some years to arrive at a new interpretation of nature against which he could grind and wield his epistemological axe; but the widespread pessimism of the early Weimar years, which, despite his personal sorrows, he did not share, gave him an immediate target at which to empty a quiver of old truths. He devoted his Leibniz Day talk of 1922, a year of runaway inflation and political assassination, to this sort of archery.

If Leibniz could have foreseen the suffering and bitter need of the German people and their science in 1922, would he not have discarded his principle of the best possible world? On the contrary, Planck answered, he would have reaffirmed it: when frustrated and disappointed by the world, Leibniz withdrew into his theodicy, "a penetrating illustration of the old truth that in the last analysis our deepest and most sacred convictions have their roots in our inner being, independent of outer experience." Already, inward-seeking physicists had found the signs of a solution to the problems of physics if not of society in an idea—the principle of least action—wondrously close to the guiding concept of Leibniz's theodicy.[92]

91. Hedwig Born to Einstein, 8 Sept. 1920, in Einstein and Born, *Briefwechsel* (1969), 58.
92. *PPA*, 42–43, 45. Cf. Planck to Wien, 1 Feb. and 13 June 1922 (AHQP).

To reach the solution, Planck continued, would require all the inner fortitude and self-discipline of the true investigator of nature—and these virtues were devaluing as fast as the mark. Planck warned against pessimism, impatience, and dilettantism, and against theosophies proferring easy answers to hard questions. He rated dilettantes and theosophists lower even than Machists; they menaced not just the philosophy but the very basis of science, "the so-called school or guild science, which amasses one grain after another in its storehouse only by hard, fatiguing, individual work."[93]

Planck's larger enemy was not the dilettantes who filled his mailbox with new schemes of the universe, but people like Oswald Spengler and Rudolf Steiner, who played on the adverse conditions of the times and encouraged a hedonistic approach to science. Spengler and Steiner blamed the ills of society on, among other things, the acquisition of technology and the loss of spirituality accompanying the triumph of classical science. Steiner, who was the greater doctor of social ills, proposed as a cure a new sort of science, a spontaneous spiritual knowing, which would deliver truths about the outer as well as the inner world, without the trouble of recourse to mathematics or experiments.

It was bad enough that the general public cultivated spiritualism and astrology and, in consequence, depreciated classical learning and its methods. It was more dangerous still, in Planck's opinion, when scientists responded, either from conviction or from a sense of expediency in assimilating with the wider culture, by condemning the methods by which their disciplines had advanced. Physicists representing the most antagonistic points of view—Stark and Einstein, Wien and Weyl, Planck and Born—associated the crisis in their theories rhetorically with the crisis in political and

93. *PPA*, 47–48.

social affairs. Some, including Weyl, Nernst, and, for a time, Erwin Schrödinger, went beyond this play on words and hinted or declared that relaxation of strict causality within their domains, as recommended by Spengler and Steiner, might well be the key to the solution of their problems.[94]

The older, influential physicists, Planck, Wien, and Einstein, spoke frequently against hedonism in science and against the proposed repeal of the law of causality. Laue took the trouble to demolish Steiner's pretensions in a firm, objective essay approved by Planck and ignored, no doubt, by its intended audience. Planck had already publicly defended positions subverted by Spengler's gospel, which, besides ruling causality out of the world, taught that all concepts, all science, must necessarily be anthropomorphic.[95] In his Leibniz Day talk of 1922, Planck challenged the fashionable by reaffirming that belief in causality is a precondition of science. At some point, he trusted, the flimsy dilettantism that tried to make do with feeling would evaporate; until then he could only try to minimize the loss of constructive effort, "which we so bitterly need today in all fields [of endeavor]."[96]

In the same address, Planck let fly at the spiritualists and theosophists for yet another fundamental failing. By unifying or, rather, mixing up science and religion, they had rendered themselves unable to give a coherent account of the relation of causality to ethical freedom.[97] The matter seemed to him so important and so menacing that he brought it again

94. Forman, *HSPS*, *3* (1971), 58–63, 77–88; Schrödinger, *Nwn*, *17* (1929), 11 (a text of 1922); Planck, *Berliner Tageblatt*, 18 June 1933, 1. Beiblatt, on crank mail.

95. Planck to Laue, 8 July 1919 (DM); Laue, *Deutsche Revue*, *47* (1922), 48–56. Cf. Forman, *HSPS*, *3* (1971), 32–34, 92–96, 101–3.

96. *PPA*, 46, 48 (1922); cf. Planck to Wien, 9 July 1922, characterizing contemporary theory as "much chaff and little wheat."

97. *PPA*, 47.

before the Academy in his next major address there, in February 1923. He advised that certain popularizers, exploiting obiter dicta of physicists, were teaching that some branches of science elude causality and so leave room for the free action of the human will. These dicta had their inspiration not only from hedonism but also, more powerfully, from positivism, against which Planck rehearsed his usual objections: positivism leads to solipsism and barrenness; science progresses by omitting the human element; physics would not exist if men had followed the candlelight of their egocentric interests rather than the blaze of the principle of causality.[98]

Planck accordingly prefaced his rebuttal to the new theorists of the human will by declaring the principle they denied. All physics and chemistry, even relativity, had been based on causality; in his view, quantum theory, too, would ultimately satisfy the principle and force a more exact formulation of it. A similar point of view informed Planck's contribution to the issue of *Naturwissenschaften* celebrating the tenth anniversary of the Bohr theory. There he pointed out that classical physics had survived the shaking it received from relativity, that Bohr's atom and all statistical physics rested on classical mechanics, and that—and this was a characteristic touch—the fact that Bohr's elucidation of spectra relied only on known fundamental constants and the natural numbers suggested that his theory would be compatible with older ideas when both had been suitably revised.[99]

Since for Planck the causal principle was coextensive with science, and science with all objective knowledge, causality must also apply to history and psychology; and he supposed that distress at its dominion over human beings prompted

98. *Wege*, 119–21, 127–34 (1923), 168 (1926).
99. *PA*, 2, 543–44 (1923). In an equally characteristic remark (ibid., 545), Planck thanked Bohr for publishing many of his important papers in German journals.

the contemporary rush to magic and spiritualism. But rightly understood—and here is his answer to theosophists, spiritualists, and positivists—the free will desired by most people is not incompatible with strict causality. Dilating a point he had made before the war, Planck resolved the apparent paradox between free will and causality by distinguishing self-examination from scientific analysis. When contemplating ourselves deliberating an action, we necessarily disturb the system under study and cannot consider it causally; we have therefore a feeling, a real and objectively justifiable feeling, of freedom of action. A well-informed external observer, however, might be able to analyze our past behavior closely enough to predict our decisions scientifically, that is, causally; and the further we are from our decisions ourselves, the closer our retrospective analysis might come to that of the omniscient externalist. But at the moment of decision we cannot proceed by causal self-analysis. How then? By Kantian imperative, by proven precepts, by a sense of duty; here intelligence must give way to character, and scientific knowledge to religious belief.

Science and religion, the latter understood as generalized ethics, must therefore be held distinct. The categorical imperative and its associated senses of duty and responsibility cannot and should not be reduced to scientific methods or principles. This is not to say that science opposes religion. To pass the threshold of science requires an act of faith similar to the acceptance of a categorical imperative or a religious doctrine, namely, a belief in strict causality. Adumbrating the theme of a later lecture, Planck concluded that science and religion can sustain no real opposition and that both are necessary: we must cultivate our religious as well as our scientific powers if we are to develop our natures fully.[100]

---

100. *Wege*, 135–41 (1923). Cf. *PhW*, 176–79, and the critical assessment in Kropp, *Zeitschrift für philosophische Forschung*, 6:3 (1952), 455.

Although Planck's reconciliation of free will and determinism had been anticipated by others, notably William James, and so was familiar to philosophers, it turns on an argument perhaps too compressed for easy understanding. Planck later set it out clearly and distinctly in his correspondence with Theodor Haering, professor of philosophy at the University of Tübingen, whose enlightenment demanded six long letters. First Planck explained that determinism makes sense only when the data on which prediction rests can be obtained without influencing or changing the system under study. "This is the basic presupposition of any sort of scientific knowledge." It does not hold for self-analysis. Hence one's will cannot be a subject of science for one's self, and the question of the will's determination makes no sense to the willing individual. Haering replied that he could not understand why, if scientific determinism holds generally, it does not apply also to "the I"; nor did he perceive why the self must be disturbed in self-examination.[101]

Planck insisted that a disturbance must occur: self-knowledge is a conscious experience; every such experience implies a change of mental state; but self-knowledge has as its object the mental state; therefore self-knowing of the knowing-self is impossible. This was not enough to complete Haering's understanding. It appeared that the philosopher had confused the "phenomenological I ('Me')" with the "actual I ('I')"; the latter can know the former, but not itself. "Therefore," Planck concluded, having reduced his argument, he thought, to the level of his grandchildren, "determinism can never invalidate freedom of the will." The philosopher still had his difficulties: he did not comprehend, he said, how observing an object could change it. Planck replied that observing does not change objects, but subjects;

101. Planck to Haering, 25 May, 6 and 26 June 1926 (Tübingen Univ. Archives).

it happens that in the case under investigation, the subject, the knowing I, is identical with the object, the willing I. That did the trick. Haering grasped the argument and Planck the difficulty of discourse with philosophers.[102] He would soon discover that quantum physicists, too, run into confusion when discussing the relationship between the knower and the known.

COPENHAGEN SPIRIT

Planck had worked intensely on the quantum theory during the war and, with the help of much correspondence with Ehrenfest, had fought to complete clarity about the statistical underpinnings and nonclassical features of the new physics.[103] At the same time he studied and then lectured on Bohr's theory of the atom as developed by Sommerfeld, with whom also he exchanged long letters on points of interpretation. That the theory had some truth was confirmed for Planck by Sommerfeld's success in marrying relativity and quantum theory and in delivering an unexpectedly healthy offspring in the form of a refined account of spectral lines.[104]

In acknowledgment of Planck's role as the originator of the approach that he had taken so far, Sommerfeld sent a gracious couplet: "You carefully cleared the land / While I gathered a little bouquet of flowers here and there"; Planck

102. Planck to Haering, 26 June, 1 July, 5 and 24 Aug. 1926 (Tübingen Univ. Archives).

103. *PAA*, 11 (1912), suggests Planck's understanding of statistics in physics just before the war; the correspondence with Ehrenfest is in ESC, the tone of which will appear from Planck to Ehrenfest, 22 Feb. 1920: "I can assure you in all seriousness that nothing is more welcome to me than a fundamental and straightforward criticism of my presentation."

104. Planck to Sommerfeld, 11 Feb. 1916, in Benz, *Sommerfeld* (1975), 97, and 12 May 1916 (AHQP/84); and to Karl Schwarzschild, 30 Mar. 1916 (AHQP/Schw), mentioning his unsuccessful attempt at an alternative theory of the Balmer series.

replied yet more graciously:[105] "What you have picked / What I have picked / We shall bind together / And since we share whatever we have / We'll make a most beautiful garland." But this garland, as Planck well knew, displayed dying classical blooms along with vigorous new shoots. He undertook to do something to minimize the contradictions and incongruities.

In the last year of the war he identified what he took to be the most important puzzle then obstructing the progress of physics: does radiation spread in space continuously, as required by Maxwell's equations, or does it propagate in the form of light quanta, as Einstein had argued? For his part, Planck told Lorentz early in 1918, he was trying to make do with the old approach, not because he ruled out the new but because Maxwell's theory had the advantage of giving definite answers. "In this way one can hope seriously to reach a contradiction with the facts, should there be one." In a lecture to the Kaiser-Wilhelm-Gesellschaft the following year, he made clear why he had come to take seriously Einstein's theory of light quanta, which he had earlier regarded as an embarrassment: Bohr's account of atomic structure had shown that light must be quantized when emitted. Does it then spread as waves or as corpuscles? "In fact, the question whether light rays themselves are quantized or whether the quantum effect occurs only in matter is the first and most serious dilemma confronting the entire quantum theory."[106] By implication, the furtherance by the Kaiser-Wilhelm-Gesellschaft of the study of the quantum theories of the director of its physics institute might pay a handsome dividend.

105. Sommerfeld, *Nwn*, *17* (1929), 481, also in Benz, *Sommerfeld* (1975), 90; the exchange occurred around the time of the first Solvay council.

106. Planck to Lorentz, 31 Mar. 1918 (LTZ/6); *PA*, *3*, 119 (1919); cf. Planck's praise of Bohr in *PA*, *2*, 543–44 (1923).

In the summer and fall of 1925, the Göttingen school of theoretical physics—Born, Werner Heisenberg, and Pascual Jordan—paid the dividend. At first Planck thought their approach to the quantum riddle hopeful. He wrote Lorentz in December that he now, "after many disappointments," expected to see the quantum theory incorporated in a unified world picture. He had anticipated that a general formulation of the new physics would replace some of the differential equations of classical theory, which imply continuity in the processes they describe, with difference equations, which might capture the characteristic jumps of quantum phenomena. The Göttingen matrix mechanics effected just such a replacement, and in a way that seemed to save the old, successful representation of interference phenomena.[107] Early in 1926 Planck brought these hopeful tidings to audiences in Karlsruhe and Berlin, embedded in a summary of his ideas about causality, statistical mechanics, and free will. However the business comes out, Planck said, in one of the worst guesses of his career, physical theory will not have to concern itself with the disturbance by the investigator of the object of his investigation, for physics excludes such complications "from the very beginning."[108]

The Göttingen school and the Copenhagen group led by Niels Bohr soon introduced precisely such complications into physics, however, to counter an initiative by Erwin Schrödinger that Heisenberg deemed "disgusting" and that Planck greeted as "wonderful" and "epoch making."[109] Schrödinger had found a way to anchor the discontinuities of the quantum theory in an old-fashioned differential equa-

---

107. Planck to Lorentz, 15 Dec. 1925 (LTZ/8).
108. *PA*, *3*, 159–62, 165–66, 170 (1926); cf. *PhW*, 194–96.
109. Planck to Wien, 22 Mar. 1926 (AHQP), to Schrödinger, 24 May 1926, in Przibram, ed., *Briefe* (1963), 6, to Lorentz, 13 June 1926 (LTZ/ 8), and to Schrödinger, 15 June 1926, in Przibram, *Briefe* (1963), 16; Heisenberg to Pauli, 8 June 1926, in Pauli, *Briefwechsel* (1979), *1*, 328.

tion for a certain wave, which he interpreted as describing the distribution of electricity within the atom. "I am reading your paper [Planck wrote Schrödinger] as an excited child listens to the answer to a riddle that has long perplexed it."[110] Planck's closest colleagues—Einstein, Lorentz, Wien, Laue—also welcomed this apparent salvation, this return to accounts of matter in motion and of the doings of electrons in space and time. They all saw difficulties, perhaps insuperable ones, in Schrödinger's interpretation, particularly in his expectation that an electron could be described in a classical manner as a superposition of waves. Waves spread; a group of them occupying a few cubic miles is not a good representation of an electron in an atom. Lorentz illustrated the difficulty by calculation; Planck thought it the main question; Schrödinger agreed, and hoped that sufficient mathematical cleverness could conquer the spreading.[111]

Alas! The Göttingen-Copenhagen group found additional difficulties and dematerialized Schrödinger's wave. Their interpretation developed from a suggestion of Born's that the wave, or rather the square of its amplitude, should be taken as a measure of the probability that one or another electron will occupy the place and state described by the wave. A wavelike probability, which, like any wave, can interfere with itself, struck Schrödinger as comic.[112] One man's joke may be another's philosophy, however, and Born's point of view that quantum mechanics offers only a calculus of probabilities became the standard interpretation.

110. Planck to Schrödinger, 2 Apr. 1926, in Przibram, *Briefe* (1963), 3.

111. For example, Lorentz to Schrödinger, 27 May 1926, and Planck to Schrödinger, 4 June 1926, in Przibram, *Briefe* (1963), 41–50, 12; Planck to Lorentz, 13 June 1926 (LTZ/8), and to Wien, 19 Dec. 1926 (AHQP).

112. Schrödinger to Planck, 4 July 1927, in Przibram, *Briefe* (1963), 18.

Planck agreed that the realistic interpretation proposed by Schrödinger could not be sustained.[113] But he emphatically refused to accept the later claim of the Göttingen-Copenhagen school that no causal theory of atomic events, no realistic objective picture of the microworld, was possible in principle.

The definitive, or Copenhagen, interpretation of quantum physics was issued in two installments in 1927. The earlier, published by Heisenberg in the spring, contained the famous uncertainty principle: it is not possible to determine to arbitrarily great exactness the simultaneous position and momentum of a subatomic particle. Heisenberg gave the example of locating an electron on a microscope stage by illuminating it with light of wavelength λ. To bring tidings of the electron, the light must be scattered by it back into the microscope's eyepiece; light so scattered must have picked up a momentum parallel to the stage in an amount $\Delta p \approx p \sin\alpha$, where 2α is the angular width of the microscope objective as seen from the center of the stage. According to the classical theory of imaging, a microscope cannot resolve points along the stage closer than a distance $\Delta q \approx \lambda / \sin\alpha$ apart. We have therefore the equation $\Delta p \Delta q \approx p\lambda$. In the classical theory this equation signifies nothing, since p and λ (both of which relate to the light) are entirely independent and each can in principle be made as small as the experimenter pleases. But in wave mechanics, $p = h / \lambda$, and we have the perplexing relation $\Delta p \Delta q \approx h$. Heisenberg explained that, by energy conservation, Δp refers to the electron as well as to the light and that in a measurement on an electron an uncontrollable, irremediable disturbance occurs, the minimum value of which is the quantum of action h. The more

113. *PA*, *2*, 637 (1927), a review of Schrödinger's *Abhandlungen zur Wellenmechanik* (Leipzig: Barth, 1927).

exactly the position is fixed, the greater the uncertainty in the momentum, and vice versa.[114]

Bohr did not like Heisenberg's representation of electrons as particles with definite, although unknowable, mechanical properties or his depreciation of the wave side of things. At the Como and Solvay meetings in the fall of 1927, Bohr presented the second and definitive installment of the Copenhagen interpretation. It stands on three main postulates: that physical theory describes the outcome of experiments and not the nature of things; that accounts of experimental setups and outcomes must be made in terms of ordinary, classical physics; and that the languages of both wave and particle theories describe all possible interactions between microentities and experimenters. The first proposition places Bohr's "complementarity," as he called his philosophy, among the epistemologies that Planck had been attacking for twenty years; it explicitly removes the properties of independently existing things from the purview of physics. The second proposition allows the analyst to communicate with his colleagues and to decide where to draw the line between the apparatus (which, of course, is composed of microentities) and the microentity under observation; in Heisenberg's microscope the apparatus includes the light, and consequently the classical result, $\Delta x \approx \lambda / \sin\alpha$, applies. The third proposition rests on experience: no known experiment had eluded elucidation in terms of classical concepts.[115]

Are not the concepts of wave (whose energy spreads in space) and particle (whose energy stays localized) antithetical? And does not light act as a wave in interference phenom-

114. Heisenberg, *Zeitschrift für Physik*, *43* (1927), 172–98. The formulation in the text is a modification of Heisenberg's original conception forced upon him by Bohr. Cf. Heilbron, *Revue d'histoire des sciences*, 200–01.

115. Bohr, in Congresso internazionale, *Atti* (1928), *2*, 565–88.

ena and as a particle in the photoeffect? Are we therefore not caught red-handed in paralogism? Not at all, according to Bohr. The experimenter decides whether to bring out the wave or the particle aspects of the experiment: the apparatus he chooses for the one cannot produce the other. For example, a double-slit experiment on light (or an electron beam) gives diffraction patterns fully describable in wave language. Particle language does not suffice to predict the patterns, but it may be made compatible with them by appeal to the relation $\Delta p \Delta q \approx h$. Let Δq refer to the distance between the slits and Δp to the associated uncontrollable momentum parallel to the plane of the slits acquired by a light quantum (or electron) in traversing them; it may therefore be clear why the beam of microentities spreads out parallel to the slits, but not why it makes the patterns it does. In short, Bohr's experimenter can never fall into contradiction in describing his results. Only counterrevolutionaries get into trouble, by putting questions of the type, "Is light (or an electron) a wave or a particle?" But that is trouble of their own making. Such questions are red herrings, according to those illuminated by the Copenhagen spirit; they do not belong to physics, which concerns itself with experiments, not with things.

Planck first heard of the "ominous relation $\Delta p \Delta q \approx h$" during his trip to Leyden in late May of 1927 to receive the Lorentz medal and to participate in an informal colloquium arranged by Ehrenfest on the necessity, if any, of interpreting elementary events statistically in quantum theory.[116] Until then Planck had continued to regard the development of quantum mechanics with equanimity and even with enthusiasm. In a lecture in mid-May to the Franklin Institute in Philadelphia, he had observed that recent discoveries did not impugn the most fundamental physical principles, like the

116. Planck to Lorentz, 6 June 1927 (LTZ/8); Ehrenfest to Planck, 8 May 1927 (ESC).

laws of thermodynamics and the equations of Maxwell, and that the wave-and-particle behavior of light had at least this to be said for it, that it presaged a conflation of corpuscular and undulatory physics. He expected that the final solution, although bringing "a profound transformation of all our physical intuitions," would satisfy his requirements: "Theoretical physics will have taken a further important step forward toward reaching its highest goal: the construction of a unified world picture."[117] He had not expected the uncertainty principle. Either it meant that physical theory depended on the skill of experimenters, which would introduce "an anthropomorphic, and therefore an unnatural element" into the world picture; or it meant that in principle it made no sense to speak of exact values of p and q, which would bring "an unacceptable limitation of the freedom of thought, and . . . a mutilation of the main instrument with which the theorist must work."[118]

The refusal of the Copenhagen school to look further, its resignation to ignorance about matters that previous physicists had believed to be capitally important, its contentment with what Einstein called "the Heisenberg-Bohr tranquilizing philosophy, or religion," astounded Planck. Quite apart from its content, the Copenhagen interpretation came with a breathless dogmatism that contrasted with his "ever-increasing aversion" to quick answers.[119] Whence this exalted pessimism, this eagerness for surrender, this "combination of resignation and enthusiasm" that, in Bohr's phrase, inspired the investigations of his followers? According to Laue, writing during the political and economic crisis of 1932, the Copenhagen interpretation was but an expression

117. *PA*, 2, 629–31 (1927).

118. Planck to Lorentz, 6 June 1927 (LTZ/8).

119. Einstein to Schrödinger, 31 May 1928, in Przibram, *Briefe* (1963), 29; Planck to Ehrenfest, 15 May 1927 (ESC, quote); Planck, *Wege*, 238–42 (1932).

of the general deep cultural pessimism that blackened the times. And that, to Laue, was just plain cowardice. "Planck has mentioned conscientiousness and loyalty as the necessary character traits of a scientist," Laue wrote. "I think that we should add patience."[120]

In 1928 the Berlin optimists welcomed Planck's successor Schrödinger, who joined them in rejecting the probabilistic interpretation as too easy a solution. He pointed to Planck's heroic struggles in the creation of the original quantum theory as an example of the patient, hopeful industry needed to bring forth healthy new physics. "I think [he wrote Planck] we are obliged to take up the same fight again from the perspectives that have lately come into view."[121] The Berlin group reaffirmed its perspective ceremonially in June and July 1929. On the earlier occasion, the Physical Society celebrated the fiftieth anniversary of Planck's doctorate by awarding him its first and Einstein its second Planck medal for outstanding contributions to physical theory. The award, a German version of the Lorentz medal, was very much a family affair, since its instigators (Born, Einstein, Laue, Schrödinger, and Sommerfeld) were among the judges and recipients. Einstein responded to his award by expressing his admiration for Planck's objectivity and his conviction that quantum physics would not stop at statistical laws and subcausality.[122]

A week later many of the same celebrants reconvened for Schrödinger's induction into the Academy, which opposed

120. Bohr to Ernest Rutherford, 3 June 1930 (AHQP); Laue, Nwn, 20 (1932), 915–16, Laue, Nwn, 22 (1934), 439–40, and Laue, Scientia, 54 (1933), 412. The events of the late 1930s naturally confirmed Laue in his opinion; Laue to C. von Weizsäcker, 21 Nov. 1942 (DM).

121. Schrödinger to Planck, 4 July 1927, in Przibram, Briefe (1963), 17.

122. Einstein, Forschungen und Fortschritte, 5 (1929), 248 (28 June 1929), in Tauber, ed., Einstein's relativity (1979), 45; Nwn, 16 (1928), 368; Pauli, Briefwechsel (1979), 2, 215.

positivism of all kinds. In his acceptance speech he put the matter at issue as follows: is quantum theory compatible with the quest, exemplified by classical mechanics, for laws that determine unambiguously the behavior of individual particles? He answered that no definitive proof either way could ever be found: the issue, rather, was which point of view—strict causality or radical probabilism—allowed the most convenient description of nature, the most useful world picture. Planck accepted Schrödinger's placing of the crux: theories must finally be judged by their fruits, by their success in providing "as true a model of nature as possible." He had no doubt that theories resting on the postulate of strict causality would ultimately be the most convenient and productive.

Planck supposed that physicists who believed that causal theories could never capture the quantum had been misled by the argument that the new mechanics ruled out in principle the prosecution of measurement to arbitrary accuracy. Hence, they inferred, by the same reasoning Planck followed in discussing the freedom of the will, the very conditions for application of a deterministic theory cannot be met. Rather than impugn causality, however, physicists should instead separate the inquiry into the conditions that define the course of a natural process unambiguously and causally from the further question whether and how far the conditions can be met experimentally. Planck located the strongest argument for the Berlin point of view in Schrödinger's work, which showed that the behavior of an atom in space and time can be completely determined if the elements of the determination are taken to be the waves satisfying Schrödinger's differential equation rather than mass points.[123]

The new situation leavened by the Copenhagen spirit brought Planck back to his old adversary Mach. He recog-

123. Schrödinger, "Antrittsrede," in PPA, 120, and Kirsten and Körber, 2, 264–66; Planck, "Erwiderung," in PPA, 121–24, and Kirsten and

nized that the discoveries of Heisenberg and Bohr about the nature of measurement might well give the impression that physics had gone astray by violating Mach's injunctions, by moving away from the description of immediate experience, by objectifying world pictures.[124] In February 1929 he returned to Leyden and to Mach, whom, at Ehrenfest's urging, he now treated more gently. It no longer was clear that positivism could bear no fruit: like it or not, quantum mechanics worked. The new physics need not, however, force anyone to abandon belief in a real outer world, although it did recommend an enlargement of the scheme Planck had proposed twenty years earlier. He began to distinguish three worlds, thus reaching a division numerically, if not substantively, the same as Mach's: where the economist saw the physical, the physiological, and the psychological, the realist put the worlds of the senses, the real, and the physicist.[125]

The physicist's world is entirely at his disposal. When this world, or world picture, appears to be stable, scientists incline to interpret it metaphysically, as asymptotically coincident with the real world; "on the other hand, in times of change and insecurity, such as we are now experiencing, positivism comes to the fore, since the conscientious scientist then inclines to draw back to the only firm starting point, the processes in the natural world." But history shows, according to Planck, that physics does not advance by oscillating

Körber, 2, 266–68 (4 July 1929); LS, 2, 58–59, 231–32. Cf. Bertrand Russell to Bohr, 30 Sept. 1935 (AHQP): "I am looking forward to seeing you for various reasons, but among others because I hope to get to understand why Heisenberg's principle is so incompatible with determinism. Left to myself I should only have thought that the things to be determined are not what used to be supposed."

124. PA, 3, 177 (1928).

125. Cf. Planck to Ehrenfest, 12 Jan. 1929, and Marga Planck to Ehrenfest, 22 Feb. 1929 (ESC); Kropp, Zeitschrift für philosophische Forschung, 6:3 (1952), 446–50; PA, 3, 181–83 (1929); and PhW, 127–31, which criticizes Planck for relaxing his intolerance of positivism.

between the two points of view: the steady removal of references to the human condition from the world picture, instanced most dramatically by the theory of relativity, continually displaces it from the sensory world toward the real.[126]

As for quantum theory, all it menaces is the mechanics of mass points: thermodynamics, the conservation of momentum, the universal constants, relativity, all survive in the world picture. Heisenberg's references to human measuring capacity have no place in the physicist's world. There, if one wishes, strict causality reigns; the appropriate subjects of causal analysis are not mass points, but Schrödinger waves. Heisenberg's uncertainty relation does inhere in some form in this world picture because it is a mathematical consequence of the characteristics of wave packets. But there can be no indeterminism, provided one does not require of the world picture information it is incompetent to give, such as the future course of an individual particle. The particle belongs to another world picture, that of classical mechanics, the language and concepts of which have mistakenly been taken over into the world of Schrödinger waves.[127]

The retention of causality would certainly weaken the tie between the new world picture and the sensory world. But in that Planck did not see only disadvantage. More radical here even than Bohr, he claimed the weakening of the bond between the world picture and experience as a mark of progress, a further step toward deanthropomorphism.[128] Although our struggle in that direction will never end, even our perpetual shortfall, rightly considered, has its advantages: "Perhaps we have every reason to consider the endlessness of this continuous struggle for the prize beckoning to us from on high as a special blessing for the questioning

126. *PA*, *3*, 183, 185–87 (1929).
127. *PA*, *3*, 202–6 (1929); repeated in several places, for example, *Wege*, 288–89 (1936), 315–18 (1937).
128. Cf. Haas, *Materiewellen* (1928), 146.

spirit of man. It insures that his two noblest impulses will endure and take fire again and again: enthusiasm and awe." This panegyric kindled Einstein's enthusiasm. At the end of a letter to Planck about the Kaiser-Wilhelm Institute for Physics, he added gratuitously: "Finally I must tell you again how wonderful I've found your discussions of positivism in connection with the modern phase of theoretical physics."[129]

With his return engagement at Leyden, Planck's original contributions to the philosophy of physics came to an end. He subsequently repeated and refined his position, after a careful study of the latest physical theory, which, in the case of Dirac's, was made more accessible to his "physical or geometrical" style of thinking by a commentary drawn up for him by Ehrenfest. In his *Positivismus und reale Aussenwelt* (*Positivism and the objective world*) (1931), his first major address to the Kaiser-Wilhelm-Gesellschaft after assuming its presidency, Planck related the inclination toward positivism and acausality to the general contemporary crisis in spiritual and material life, slashed out as usual against positivism, and insisted that no objective reason compelled renunciation of strict causality in the world picture. The quantum limits measuring, to be sure, but not the knowable; "the results of measurements must be extended through free speculation." Against all pessimists, quantum or otherwise, he affirmed the power of the human intellect, the operation of a limitless apparatus, "that is, the flight of our thoughts."[130] In *Der Kausalbegriff in der Physik* (*The concept of causality in physics*) (1932), Planck criticized indeterminists for the old positivist sin of confusing the sense world with the world picture and for concluding that phenomena could compel the abandonment of causality.[131]

129. *PA*, *3*, 208 (1929); Einstein to Planck, 15 Nov. 1930 (AE).

130. Ehrenfest to Planck, 1 June 1927, and Planck to Ehrenfest, 24 July 1927 (ESC); Planck to Sommerfeld, 2 Feb. 1929 (AHQP/84); *Wege*, 201–8, 213 (quote) (1930), 318–19 (quote) (1937). Cf. *PhW*, 167.

131. *PA*, *3*, 225–30 (1932); cf. ibid., *3*, 247 (1933); LS, *2*, 315–16.

IRRATIONAL RELATIONS

In the early 1930s German science appeared more likely to confirm the pessimists than to vindicate Planck's confidence in the flight of human thought. Economic depression then cut the income of the major supporters of world-class science in Germany, the Notgemeinschaft and the Kaiser-Wilhelm-Gesellschaft, far below what they deemed necessary to sustain their work.[132] Political extremism squeezed out moderate parties like the Deutsche Volkspartei and brought to leadership men unreachable by reasoned argument. Overt and vicious anti-Semitism reached deep into academic life and elevated Aryans to posts that, by merit, should have gone to Jews. The general public opposed science and science-based technology as primary causes of overproduction and underemployment. The government promoted applied science at the expense of pure science as the quickest way out of economic and military inferiority. And in theoretical physics, the purest of pure sciences and Planck's own life's work, there was a preference for easy solutions, an air of resignation, a loss of direction.

Although Planck could not combat the causes of the crisis, he considered it his duty to speak against the misconstrual of the basis of physics, not only for the salvation of his subject but also for the well-being of all thinking people, whose worldview might easily be subverted by cracks in the foundations of physics. In his reply to Schrödinger's acceptance speech at the Academy, he had already stressed that if physics did not resolve the question of causality satisfactorily, "truly dangerous" consequences might follow far beyond the limits of science.[133] And by the early 1930s he could point to writings of Pascual Jordan, who tried to base a radical free-

132. The Notgemeinschaft's budget dropped from 8 million marks (897,000 for physics) in 1928 to 4.4 million (385,000 for physics) in 1932, according to Richter, *Forschungsförderung* (1972), 24.

133. Planck, in Kirsten and Körber, *2*, 266 (1929).

dom of the will, vitalism, and extrasensory perception on an acausality sanctioned by quantum physics. Jordan brought the clout of physics to the support of the depth psychology of Freud, and Wolfgang Pauli did the same for Jung. As Planck foresaw, theologians grasped at the uncertainty principle, elucidated in the style of Heisenberg, as evidence of indeterminism, and they saw indeterminism as ground or room for personal freedom and moral responsibility. He may also have envisaged the political exploitation of the Copenhagen spirit, vitalistically misinterpreted, in the interest of irrational political doctrines. An influential theologian, Bernard Bavink, pointed out the agreement between Hitler's notions and policies, on the one hand, and the antimaterialist, which is to say idealist, vitalist, and anticommunist, consequences of the new physics, on the other. The notion for a time pleased the editors of the leading Nazi newspaper, the *Völkischer Beobachter*, which recommended Bavink's *Naturwissenschaft auf dem Wege der Religion* (*Science on the path of religion*) to its readers until Lenard apprised them that Bavink favored Einstein and had the endorsement of Planck.[134] The physicist most responsible for spreading a vitalistic interpretation of Bohr's philosophy, Jordan, became a Nazi.[135]

In *Positivismus und reale Aussenwelt*, in a widely circulated lecture to the Verein deutscher Ingenieure (Association of German Engineers) in 1933, in an interview in the *Berliner Tageblatt* for the same year, and again and again in the late 1930s, Planck emphatically rejected facile inferences from the formalism of quantum physics to problems in bi-

134. Bavink, *Unsere Welt*, 25 (1933), 233; Heilbron, *Revue d'histoire des sciences* (in press); *Völkischer Beobachter*, 15 Dec. 1933 and 15 Jan. 1934, and Lenard to Alfred Rosenberg, 9 Jan. 1936, in Poliakov and Wulf, eds., *Das Dritte Reich* (1959), 296. Planck admired Bavink: Planck to Hildegard Gravemann, 4 Jan. 1943 (MPG).

135. Jordan to Bohr, May 1945 (AHQP).

ology and psychology. He objected particularly to moving from acausality to free will. Quite apart from the question of the legitimacy of acausality in physics, its establishment as the basis of behavior would replace deliberation with blind chance. Free will as justification of the moral responsibility of the individual would not thereby be saved: rather, every act, however savage, might be excused as a spontaneous manifestation of an uncontrollable, unpredictable, unfathomable chance.[136] The introduction of radical acausality into the physical world picture could eventuate in, or give support to, concepts with socially catastrophic consequences.

What goes into the world picture is a matter of choice, and what we choose is a matter of faith: logic cannot compel belief in a real outer world or in a preestablished harmony between thoughts and things or in an asymptotic coincidence of the world picture and the real world.[137] The final basis, the last authority, for the fundamental principles of physical science lie deep in the individual. Planck wrote Bohr in 1930, after returning from reading *Positivism and the objective world* to the Danish Physical Society: "The highest court is in the end one's own conscience and conviction—that goes for you and for Einstein and every other physicist—and before any science there is first of all belief. For me, it is belief in a complete lawfulness in everything that happens."[138]

The need to choose fundamental principles by appeal to

136. *Wege*, 215–18 (1930), 223–42 (1932); *PA, 3,* 250 (1933); cf. *PhW*, 197, 200–202. The pervasive misunderstanding of Planck's position may be estimated from the article "Wille" in the *Lexikon für Theologie und Kirche* (1938), *10*, col. 915, which cites him as authority for the view that the concept of causality in modern physics leaves room for free will.

137. *PA, 3,* 239 (1932); *PPA*, 152 (1935).

138. Planck to Bohr, 19 Oct. 1930 (AHQP); Planck's lecture appeared in Danish in *Fysisk tidsskrift, 28* (1930), 161–81.

something other than logic introduces something nonscientific at the base of science. Adopting a favorite word of Bohr's, Planck characterized this something as "irrational." The word would not have occurred to him in connection with the foundations of science in the old days, when he exposed Mach's fallacies to the bright light of reason. But there remained from the combat with Mach the touchstone of fertility, which Planck now applied to social as well as to scientific fruits. And here, as he told the Association of German Engineers, his optimistic belief in progress toward the true and real held the advantage over positivistic pessimism, for it inspired people to work. "In science, as in all other fields of cultural development, work is the only sure way to health and success. As long as our German people are at work, but only then, we need not doubt a better future despite all the miseries of the present."[139]

Assigning responsibility for choosing among irrational foundations for science in accordance with one's conscience and conviction aggravated the tension mentioned earlier in Planck's method of deanthropomorphizing world pictures. (Planck had taught that the creativity of artistic human personalities had largely expunged the human element from physical theory.) The tension increased in step with the gravity of the problems confronting German physics. In imperial times Planck spoke of the creative act and fancy of the individual theorist merely to humanize physicists in the public eye by assimilating them to artists. Toward the end of the war and in the early Weimar years, he referred the special insights of the individual to elements of his national culture, in order to ameliorate the psychological effects of the isolation of German science: "The history of international science has shown again and again that science, just as art and religion, can prosper in the first instance only on national soil. Only on this basis is a fruitful association of peoples in

139. *PA*, *3*, 253–54 (1933).

honorable competition possible."[140] The same thought occurs in his writings during the crisis around 1930: although science is international, scientists draw their culture from their country; if the culture is rich enough, they can, to a large extent, do without the rest of the world; fortunately, as the then recent developments in theoretical physics showed, Germany still had what it took.[141]

After 1930 Planck expanded the contributions of the individual beyond cleverness and cultured imagination to include conscience and personal values, in order to combat the tendencies of the new positivistic world picture that he thought obnoxious both to science and society. In a word, as he wrote Bjerknes: "The scientific and the purely human . . . cannot be divided." Einstein judged this doctrine to be a protective device, the basis of a comfortable, imaginary, controllable world. "Within this picture [Planck] places the center of gravity of his own soul, so that he will find in it the rest and equilibrium which he cannot find within the narrow circle of his restless personal relations to everyday life."[142] Planck's final world picture was not the severe, transcendental, objective, dehumanized construction that had attracted him in his early career, but a place of retreat from the troubles of science and society. It was also congenial to the Berlin theorists and to Sommerfeld.[143]

Apart from a few philosophers who dismissed Planck's

140. *Sb* (1926), lxxxv–lxxxvi (also in *PPA*, 94), following the lead of *Sb* (1917), 37–38. Cf. Forman, *Isis*, *64* (1973), 158–60.

141. *PA*, *3*, 209 (1930); *PPA*, 131 (1930), echoed by Schreiber in Abb, ed., *Aus fünfzig Jahren* (1930), 9, 12 (a festschrift for Schmidt-Ott). Planck made an exception for transcendent genius and in the cause of international good manners in his contribution to an English festschrift on the centennial of James Clerk Maxwell's birth; *PA*, *3*, 352 (1931).

142. Planck to Vilhelm Bjerknes, 30 Dec. 1932 (Oslo Univ. Library); Einstein, in Planck, *New science* (1959), ix. Cf. Heisenberg, *Nwn*, *23* (1935), 321.

143. Sommerfeld to Moritz Schlick, 17 Oct. 1932, quoted in Pauli, *Briefwechsel* (1979), *2*, 58; Benz, *Sommerfeld* (1975), 157.

resolution of the conflict between free will and determinism as professionally inadequate,[144] the exponents of the Copenhagen spirit and the spokes of the Vienna Circle (the most influential school of the positivist philosophy of science) were the most evident opponents of his teachings. In the view of the Vienna Circle, as stated by the physicist-philosopher Philipp Frank, Planck was the leading representative of the standard school metaphysics, according to which an objective world exists, full of things ready to be discovered, like America waiting for Columbus. Frank objected that Planck had supposed too many worlds: where positivists made do with two, the sensory and the mathematical, Planck required a third, a superfluous "Drittes Reich" (in Frank's witty slur) of real things.[145] As Jordan, who fancied himself a follower of Mach and the Vienna Circle, explained the position: positivists do not deny the existence of a real outer world, they deny that the claim that it exists has any meaning. For his part, he had no trouble rejecting Planck's third world on the authority of quantum mechanics, which, he said, involved the assumption that a microsystem has no definite characteristics before a measurement is made upon it. There is no nonarbitrary criterion for distinguishing system from measurer, inner from outer, individual from social.[146]

Heisenberg strove for a balanced view. He judged *Positi-*

144. For example, Groos, *Determinismus* (1931), 141; Davidson, *Free will* ([1937]), 82; Erickson, *Journal of philosophy*, *35* (1938), 208–11; but cf. the appreciative review of *Kausalbegriff* by Ferguson, *Nature*, *130* (1932), 48.

145. Frank, *Erkenntnis*, *1* (1930–31), 126; Frank, *Erkenntnis*, *5* (1935), 71; Frank to Bohr, 9 Jan. 1936 (AHQP).

146. Jordan, *Nwn*, *22* (1934), 488–89, and Jordan, *Anschauliche Quantentheorie* (1936), 302–5, 308, 317–18. Jordan's conceptions about biology, vitalism, free will, etc., are hardly positivistic; Planck tried to make him see that even in his physics he admitted something like a real outer world. Planck to Jordan, 7 July 1936 and 21 Mar. 1937 (Jordan Papers, SPK); Planck, *Nwn*, *28* (1940), 779.

vismus und reale Aussenwelt to be "most probably a lost cause and consequently of little practical value," but he applauded Planck's recognition that science necessarily has an irrational taint. Of course he disapproved of Planck's particular irrationality, the supposition of an objective world run by strict causality. Heisenberg ascribed this irrationality to Planck's "religious-moral conception of life . . . , which enables him to walk a straight and almost too certain a path even where immeasurable epistemological abysses threaten on either side." Here Heisenberg was no doubt correct and—except for the insinuation that Planck had not come to grips with the difficulties—almost laudatory.[147] Perhaps Heisenberg's winning of the Planck medal for 1933 (on Planck's nomination) had caused him to pull his punch.[148]

That was Pauli's interpretation. He prayed for the spiritual health of his straying brother Heisenberg. "May the spirit that dominates Planck's scientific production and personal life not take over your publications and your life too strongly." And he severely criticized the favorable remarks in Heisenberg's review. "If you grant that statements about 'the reality of the outer world' make sense, you give the devil of 'ism-philosophy' your little finger, and soon he will take the whole hand."[149]

Among people who were not quantum theorists, however, among philosophers and theologians in Germany and elsewhere, Planck's writings were major sources for the content and interpretation of modern physics. Random entries from

147. Heisenberg, *Zeitschrift für angewandte Chemie*, *44* (1931), 189, and Heisenberg, *Nwn*, *21* (1933), 608 (quote).

148. Planck to unnamed correspondent, 15 Jan. 1933 (Autogr. I/287, SPK).

149. Pauli to Heisenberg, 29 Sept. 1933, in Pauli, *Briefwechsel* (1979), 2, 214. Pauli thought Planck's "mind [Geist]" to be very "sloppy [schlampig]"; Heisenberg understood the judgment to refer to Planck's epistemology, not to his politics (letter of 7 Oct. 1933, ibid., 218); cf. the editor's remarks, ibid., 211, 215.

a very long list include the biologists Paul Jensen and Max Hartmann; the philosophers Albert Eagle, Helmut Groos, and J. W. A. Hickson; the theologians William Ralph Inge and Rudolf Keussen.[150] All accepted Planck's point of view. A striking example of his authority for the interpretation of physics to nonphysicists occurs in a dictionary of philosophical concepts published in 1944. Planck appears as the primary reference not only for "Determinism and Indeterminism" and "Science and Life," but also for "Heisenberg Uncertainty Relation."[151]

150. Jensen, *Erkenntnis*, *4:3* (1934), 181–82; Max Hartmann, *Vorträge und Aufsätze* (1956), 2, 111–13, 116; Eagle, *Religion versus science* (1935), 243–48; Groos, *Determinismus* (1931); J. W. A. Hickson, *Philosophical review*, *47* (1938), 601; Inge, *God and the astronomers* (1933), 58–64; Keussen, *Willensfreiheit* (1935), 27–37.

151. Hoffmeister, *Wörterbuch* (1944), s.v. "Determinismus und Indeterminismus," "Wissenschaft und Lebenspraxis," and "Heisenbergische Unsicherheitsrelation."

| 4 |

IN SHIPWRECK

On 30 January 1933, when Hitler became Reich chancellor, Planck held two key positions in the scientific establishment of Germany. As secretary of the Academy and president of the Kaiser-Wilhelm-Gesellschaft, both of which depended on the Reich for most of their support, he had to work with the new regime. And the regime had something to gain from Planck: the identification with the new Germany of a patriot of unblemished character and international reputation, whose racial stock admitted no doubt. Planck recognized that he was well placed to strike compromises for the sake of science, and at first he had reason to believe that his maneuvering, however distasteful to himself, alleviated the sufferings of some colleagues and preserved important elements of their institutions.

Also at first, and along with many others, he believed that Hitler would have to moderate Nazi policies in order to stay in office. This is a capital point in judging the behavior of Planck and men like him. They hoped that what they deemed to be the valuable contributions of National Socialism—the call to national cultural renewal, unity, and glory

149

—would subsist, and they expected that the excesses would fade away.[1] Planck had compromised fruitfully in preserving and even furthering science and science policy during the Weimar years. He might have hoped, as Schmidt-Ott did, that science could make its case more easily to the führer, "who has succeeded in rescuing us from a century of miserable party politics and in forging us together as a people," than to the dozen political parties that had perplexed the late Republic.[2]

The policy of old men like Schmidt-Ott and Planck became one of salvage, an effort to protect science without offending a regime whose "Weltanschauung," as Planck labeled it to Einstein, they did not understand. They complied openly in small things and did not protest publicly against great injustices, they accommodated to insure that lesser men did not take their places, and they strove to persuade younger colleagues to steer a similar course. Planck's policy and his justification of it appear in his advice to Hahn, Heisenberg, and Schrödinger. To Hahn, who had proposed that thirty prominent German professors make a declaration in favor of their Jewish colleagues, Planck replied, as Hahn remembered: "If you bring together 30 such men today, then tomorrow 150 will come to denounce them because they want to take their places." To Heisenberg, who debated whether to resign and emigrate, Planck said that the gesture would accomplish nothing and that a higher service would be performed by remaining and guiding young people who would "constitute so many seed crystals" for the eventual revival of scientific research in Germany. As for Schrödinger, who, although not Jewish, insisted on returning to Austria, Planck begged him to take a leave of absence, a step that would have the dual advantage of raising a hope for his even-

1. Cf. Glum, *Wissenschaft, Wirtschaft und Politik* (1964), 436.
2. Schmidt-Ott, NDW, *Berichte, 12* (1932–33), 12. Cf. HH, 70; ADB, 58–60.

tual return to Berlin and avoiding a protest resignation damaging to the colleagues he would leave behind.[3]

During the first days of Nazi power, Planck considered leaving his posts. Age recommended, but duty denied, resignation. The forces that defined his duty and bound him to it appear from correspondence with Ehrenfest, who turned to him for psychological support. Ehrenfest was clearsighted and warmhearted toward others, but overhumble and uncharitable to himself; and the spectacle of irrationalism rampant in Germany combined with family problems to deepen the depression to which he was prone. He saw in Planck a tower of strength, "a man of immensely noble purity," in whose very glance was reassurance. Ehrenfest wrote: "In this time of ever-increasing constraint and *complete* doubt, it means very much to me that your eyes always look at me with a warm and peaceful attentiveness—it means more than I can say without being effusive."[4] It was this sort of appeal and dependence and the possibility of easing hardships that kept Planck in office. Marga Planck wrote Ehrenfest about her husband's actions: "He has often wished that he could withdraw from official matters and great responsibility." But he could not: "Now everyone counts on his help." The political circumstances demanded his entire, diminishing strength.[5]

Confirmation that Planck's sense of responsibility di-

3. Hahn in MPG, *Mitteilungen* (1957), 244, and Hahn, *My life* (1970), 140; Heisenberg, *Physics and beyond* (1971), 151–52 (a Thucydidean reconstruction); Planck to Schrödinger, 19 Nov. 1933 (AHQP/92), and to Laue, 11 Sept. 1933 (DM): if Schrödinger must resign, he should do so quietly and allege reasons of health. Cf. ADB, 209, 227; Planck to Einstein, 13 Apr. 1933, in AH, 78, 83, 84.

4. Ehrenfest to Planck, 26 Mar. 1933 (ESC); the remark about purity is handwritten in the margin of Ehrenfest's carbon copy.

5. Marga Planck to Ehrenfest, 26 Apr. 1933 (ESC); Planck to Anton Kippenberg, 12 May 1933 (KA). Planck could not help Ehrenfest, who died by his own hand soon after writing Marga.

rected his actions, and not his desire to retain power or any sympathy for the new government, comes from a report from Niels Bohr's brother Harald, a mathematician who was in Berlin in May 1933. "Planck was as always extremely nice and anxious to explain [that] his reason of staying in his position as director of the Kaiser William Institute [*sic*] was to try to do all in his power to help in the situation, not only in this institute, but also in the different universities, and in fact I understand that he is rather the only person, who may have some success in his efforts."[6] These efforts did not include protests or manifestos. Here the bitter experience of the Appeal of the Ninety-three Intellectuals reinforced a policy in consonance with Planck's age and style. He had written Paul Langevin some years earlier, in connection with a proposed declaration in favor of international cooperation, that he knew of no manifesto that had achieved its purpose. "Usually, I've seen exactly the contrary. Inconsistency inevitably occurs, the declaration is misunderstood, falsely interpreted, and used to slander its signers."[7] He worked behind the scenes, helping some people dismissed from state jobs to find private ones, assisting emigration when necessary and advising against it when possible, keeping his institutions going.[8]

He thought, rightly no doubt after the first months of Nazi rule, that the less said about his endeavors the more effective they would be. When the *New York Times* commended his speech at the twenty-fifth anniversary meeting of the Kaiser-Wilhelm-Gesellschaft in 1936 ("Max Planck, to his everlasting glory, went as far as common sense per-

6. Harald Bohr to R. G. D. Richardson, 30 May 1933 (Richardson Papers, Library of Congress).
7. Planck to Paul Langevin, 24 Sept. 1925, in André Langevin, *Paul Langevin* (1971), 123–24.
8. For example, Planck to Laue, 11 Sept. 1933, re Lise Meitner; 31 Mar. 1934, re Richard Courant; and 21 Aug. 1935, re Arnold Berliner (DM); to Schrödinger, 10 Nov. 1933 (AHQP).

mitted in defending the original policies and reiterating his position that personalities and brains count for more in scientific research than race or totalitarianism"), Planck worried that the notice would accomplish exactly what he wished to avoid, "namely, drawing official attention to men like [Otto] Meyerhof and [Otto] Warburg," institute directors who happened to be Jewish.[9]

Planck's most notable move behind the scenes was an interview with Hitler, which probably took place in May 1933. Planck hoped to convince the führer that the forced emigration of Jews would kill German science and that Jews could be good Germans. According to Planck's recollection, Hitler replied that he had nothing against the Jews, only against communists, and then flew into a rage. Contemporary reports of the interview circulated widely. One, which confirms Planck's recollection, quoted Hitler as saying, "But we don't have anything against the Jews, on the contrary we protect them." According to the informant, Planck considered this "the worst possible reply he could get, because it took [from] him every basis for further negotiation."[10]

Heisenberg spoke with Planck shortly after the interview and interpreted Hitler's declarations according to their literal meaning. Therefore, in service of the policy of stay and save, he urged Max Born and James Franck, both Jews, to take advantage of certain provisions of the law "cleansing" the civil service that allowed them to retain their professorships. They had no cause for fear, Heisenberg wrote to Born:[11]

9. "The last stand," *New York Times*, 12 Jan. 1936; Planck to Laue, 31 Jan. 1936, and 24 Nov. 1941, the latter lamenting Meyerhof's emigration (DM).

10. Planck, *PB, 3* (1947), 143; Charlotte Schoenberg to Richardson, 27 July 1933 (Richardson Papers, Library of Congress).

11. Heisenberg to Born, 2 June 1933 (quote), in Pauli, *Briefwechsel* (1979), 2, 168, and to Niels Bohr, 30 June 1933, in ibid., 204. Sommerfeld saw more clearly and expected that Heisenberg too might soon want to leave; marginal note on W. L. Bragg to Sommerfeld, 29 May 1933 (Bethe/3, Cornell Univ. Library).

Planck has spoken . . . with the head of the regime and received the assurance that the government will do nothing beyond the new civil service law that could hurt our science. Since on the other hand only the very least are affected by the law—you and Franck certainly not, nor [Richard] Courant—the political revolution could take place without any damage to Göttingen physics. . . . In spite of [some dismissals], I know that among those in charge in the new political situation there are men for whose sake it is worth sticking it out. Certainly in the course of time the splendid things will separate from the hateful.

Born and Franck and also Courant ignored this ill-considered advice and left their institutes along with their Jewish subordinates who had no shelter under the new law. The garden of German mathematical physics immediately turned into a desert—and the exemplar of the effectiveness of Nazi science policy.

Heisenberg soon learned how dangerous it is to trust to the words of princes, especially words got at second hand and invested with the hopes of the hearer. By 1936 he and Planck had come under threatening personal attack by Stark and his allies in the government, and the Academy and the Kaiser-Wilhelm-Gesellschaft faced dictated reorganization. Over seventy-five years of age, Planck still stuck to his posts, which he relinquished only during forced reconstructions in 1937 and 1938. Then, at the age of eighty, he went about the country making speeches; eventually he preached in occupied territory on the cultural values of science and its complementary relation to religion. These trips could not have been arranged without the concurrence of government and military officials, not all of whom, however, approved his sermons. Was the spokesman a patriot? Retrospective judgment has been at least as difficult, although not as painful, as Planck's own decision making. At one pole, East German commentators think that he behaved creditably, even heroically; at the other, judges in the West, especially refugees or their historians, criticize his refusal to speak out or resign and condemn his apparent acquiescence in unsavory actions forced upon him by his official capacities and his policy of accommodation.[12]

As Captain

When the Nazis came to power, Einstein was in the United States. On 10 March 1933 he made public his decision not to return to Germany, which, he said, no longer enjoyed "civil liberty, tolerance, and equality of citizens before the law."[13] The hate campaign against him heated up in the rightist press, which now could add slanderer of the German people to its stale charge of subverter of German culture. The reports of Einstein's statements much distressed Planck, who had a deep affection for Einstein and continued to do him little services at the Academy up to the final break.[14] Not only did the reports make it difficult to defend him, they also, in Planck's opinion, provoked the reverse of what Einstein intended. Planck wrote: "By your efforts your racial and religious brethren will not get relief from their situation, which is already difficult enough, but rather they will be pressed the more."[15] He reminded Einstein that the value

12. For example, *PhW*, 226 (East); Frank, *Einstein* (1947), 236 (West).

13. Einstein, interview with *New York World Telegram*, in Clark, *Einstein* (1971), 458, 462.

14. For example, looking into the bureaucratic consequences of Einstein's wish to resign his German citizenship and helping with details about travel leaves. Einstein to Planck, 17 July 1931; Planck to Einstein, 20 Jan., 4 and 6 Mar. 1933 (AE).

15. Planck to Einstein, 19 Mar. 1933 (AE), in AH, 78. The same view was held by opposite sides, for example, Stark, *Nature*, *133* (21 Apr.

of an act lies not in the motives behind it, but in its consequences.

This doctrine, unlike a categorical imperative, is easier to announce than apply. What count as consequences? When should the accounting be made? In the case of Mach's epistemology, Planck left out of the reckoning the inspiration that he and Einstein had derived from the false teachings of the Economist of Science, and so managed to condemn them as valueless. In the case of Einstein's advertisements of Nazi insanity, Planck judged that the immediate effects on German Jews would be deleterious, whereas a retrospective assessor might find that much wider protests, although perhaps more injurious than silence in the short run, would have brought world opinion more effectively to bear or encouraged the resettlement of more Jews earlier.

In 1933, after consulting with Planck, the presiding secretary of the Academy, the meteorologist Heinrich von Ficker, asked Einstein to explain his conduct. Receiving no answer, Ficker consulted American newspapers reporting Einstein's remarks and concluded that the Academy would have to eject its tarnished ornament. He asked Planck, who was about to leave for vacation in Italy, to urge Einstein to resign. Planck complied, convinced that Einstein's criticism of the Reich abroad effectively barred him from state offices at home and afraid that a formal proceeding for expulsion would place Einstein's friends in a most severe conflict of conscience. "For although a deep chasm separates me from him in political matters, yet I am completely certain that in the history of future centuries the name of Einstein will be

1934), 614 ("The withholding of criticism of the new regime in Germany, or at least a conscientious regard for truth in scientific circles, will be to the advantage not only of international cooperation but also of the Jewish scientists themselves"), and Otto Warburg, according to Allan Gregg, diary, 25 Oct. 1933 (RF 717/1).

celebrated as one of the most brilliant stars that have shone in our Academy."[16]

Einstein had already sent in his resignation. The Nazi minister in charge of educational and cultural matters, Bernhard Rust, refused to allow the matter to die there and ordered the Academy to make a public explanation. A press release, drawn up hastily on 1 April by the only secretary of the Academy then in Berlin, the orientalist Ernst Heymann, charged that Einstein had participated in a slander campaign against Germany. The statement appeared on a day otherwise sufficiently full of tension: good Germans boycotted Jewish businesses, and Hitler's brownshirts removed Jewish students from the libraries and laboratories of the University.[17] It is said that the ministry had hoped to add to the day's enjoyments an announcement of Einstein's expulsion from the Academy and received his resignation, which preempted its plan, with "indescribable rage."[18]

Heymann's declaration ended in a most offensive paragraph, which exposed, among other things, the fond academic fiction about the separation of science and politics: "The Prussian Academy of Sciences is the more sensitive to Einstein's work of agitation abroad since it and its members have felt themselves bound closely to the Prussian state from remote times and, while observing due strict restraint in political questions, has firmly supported the national purpose. For this reason it has no cause to regret Einstein's resigna-

16. Ficker to Einstein, 18 Mar. 1933, to Auswärtiges Amt, 29 Mar. 1933, to Planck, 29 Mar. 1933, and Planck to Ficker, 31 Mar. 1933, all in Kirsten and Treder, *1*, 243–45; Planck to Einstein, 31 Mar. 1933, in AH, 78. Cf. LS, *3*, 7–11, Hermann, *New physics* (1979), 75–76, and Planck to Einstein, 18 June 1933 (AE), re Einstein's resignation from the order *Pour le mérite*, of which Planck was the chancellor.

17. Einstein to AW, 28 Mar. 1933, and Ficker to Planck, 8 Apr. 1933, in Kirsten and Treder, *1*, 246, 253; cf. ibid., 249–51, 257, 259.

18. Laue to Ficker, 11 July 1947, in ibid., 253.

tion."[19] The last phrase outraged Laue and Paschen, who forced a meeting to discuss retraction; but the meeting endorsed Heymann's action, thanked him, and took note that, under the circumstances, he rightly gave precedence not to Einstein's scientific standing, but to "the advice of men experienced in foreign affairs." Even Paschen voted for the resolution. It was, Laue later said, "one of the most appalling experiences of my life."[20]

The matter still did not die. In accepting Einstein's resignation, Ficker had expanded the Academy's charges: not only should a good German who had belonged to the Academy for so long have refrained from criticizing the new government, he should have defended it against the calumnies circulating about it abroad. Planck retrospectively approved Ficker's letter as "written in a thoroughly dignified and elegant manner." Still, the Academy had acted only on the knowledge of newspaper reports; its handling of the case, Planck feared rightly, would not contribute to its reputation. Einstein protested that he had participated in no campaign of persecution, that the official pronouncements of the Nazi government gave enough scope for adverse commentary, that Germany was the victim of a mass psychosis, and that the Academy had joined the slander against him without bothering to examine the evidence.[21] Unrepentant, he added that a defense of Germany would have been a betrayal of the true values of German culture. "By bearing such a witness under current conditions, I would have contributed—if only indirectly—to the brutalization of morals and the destruction of all contemporary civilization."[22]

19. Ibid., 248, 257–61.
20. Laue to Ficker, 11 July 1947, in ibid., 274.
21. Ficker to Einstein, 7 Apr. 1933, Planck to Ficker, 13 Apr. 1933, and Einstein to AW, 10 Apr. 1933, in ibid., 252–55.
22. Einstein to AW, 12 Apr. 1933, in ibid., 262. Cf. Clark, *Einstein* (1971), 467–71.

When it came Planck's turn to preside, he did what he could to soften posterity's judgment. He wrote into the minutes of the Academy's meeting of 11 May 1933 that the overwhelming majority of German physicists realized that Einstein's work could be compared in importance only with that of Kepler or Newton. "Therefore it is . . . deeply to be regretted that Einstein has by his own political behavior made his continuation in the Academy impossible."[23] Einstein had written Planck recalling the years of persecution he had suffered, almost in silence. "But now the war of annihilation against my defenseless Jewish brothers has forced me to place whatever influence I have in the world on their side of the balance." He held nothing against Planck, whose profession of continuing friendship he warmly, although only briefly, reciprocated.[24]

Later in the year Planck proposed Laue as Einstein's successor in the Academy's nonteaching professorship. His fellow secretaries agreed and submitted the proposal to the responsible ministry, which asked Lenard and Stark, now reaping their reward for backing Hitler, for advice. Lenard observed that Planck, "politically so worthless a character," had brought Einstein into the Academy; that Laue was Einstein's closest friend; that Laue had just compared, in public, the Church's persecution of Galileo with the Nazi's treatment of relativity; and that neither Planck nor Laue should have any further opportunity to mislead German youth. Stark wrote as president of the Reichsanstalt, where he had been installed since May: "If Planck and von Laue retain influence, it will have a worse effect than if Einstein himself were there." He gave as final proof of Laue's complicity the observation that the comparison between the Nazis and the

23. AW, "Protokoll," 11 May 1933, in Kirsten and Treder, *1*, 267. Cf. LS, *3*, 170–71.

24. Einstein to Planck, 6 Apr. 1933, answering Planck to Einstein, 31 Mar. 1933, in AH, 79.

persecutors of Galileo had received "the enthusiastic applause of all the Jews and their fellow travelers present."[25]

The fight over Laue's promotion was one episode in an important test of strength between the Reich and the Academy. In furtherance of his plan to become the führer of German science, Stark had brought government pressure to have himself made an academician. On 30 November 1933 a proposal to that effect was put forward by Karl Willy Wagner, Planck, and—of all people—Paschen, who drew it up. It was a case of preservation by compromise. Paschen wrote Ficker that opposition would be "tactically a false step and even dangerous." Unmanned by the attack on his stewardship (Stark had publicly condemned the Reichsanstalt's products under Paschen as "crap"), he meekly submitted to what he thought necessary for the good of the Academy: "Now is indeed the time for sacrifice."[26] As for Planck, he had got some idea of what might come to those who defied the orders of the Reich from his interview with Hitler. But Laue, understanding perfectly that the admission of Stark would make the Academy a cipher at home and a laughing-stock abroad and that it would further strengthen a hand already strangling German physics, led an opposition that succeeded in tabling the proposal. That was on 14 December. Two days later Stark fired Laue from the consultancy at the Reichsanstalt to which Nernst had appointed him in 1923.[27] Ficker then circulated reports of Stark's insane attacks on the scientific work of Academy members Planck,

25. Planck et al., "Antrag an das Kultusministerium," 30 Nov. 1933, Lenard to Kultusministerium, 22 Dec. 1933, and Stark to same, 9 Jan. 1934, all in Kirsten and Treder, *1*, 267–69; Laue, *Physikalische Zeitschrift*, *34* (1933), 889–90; Planck to Laue, 11 Sept. 1933 (DM).

26. Paschen to Ficker, 10 Nov. 1933, in LS, *3*, 173; cf. Paschen to Bohr, 1 Feb. 1935 (AHQP).

27. D. Hoffman, AW, ITGOW, *Kolloquien*, *35* (1984), 16.

Schrödinger, Haber, and Laue, and on 11 January 1934 the proposers withdrew their nomination.

Ficker considered Laue's opposition to Stark as natural, by which he meant subjective, personal, and political, and Planck's final position, which decided the affair, as balanced and justified. "I could suppose that he [Planck] was objective enough not to let himself be prejudiced completely against Stark by Stark's attacks on theoretical physicists."[28] From which it appears that meteorologist Ficker had not accurately gauged the temperature of the situation, nor had he understood that the pressure that Stark and Lenard were exerting against theoretical physics had long since broken the bounds of academic dispute.

Laue's courage may have emboldened Planck to more open defiance in another matter, which, to be sure, had only symbolic value. On 30 April 1933 Fritz Haber informed Rust that he would not make use of the exemption that allowed Jews who had served in the First World War to retain their state offices. He could not continue as head of the Kaiser-Wilhelm Institute for Physical Chemistry, which he had made a glory of German science, if required to follow Nazi "tradition." "My tradition [he wrote Rust] requires of me that in my scientific position I consider only the professional accomplishments and character of applicants when I choose my co-workers, without asking about their racial makeup."[29]

Haber, whose devotion to Germany had brought him to chemical warfare, fled to England. Planck, as president of the Kaiser-Wilhelm-Gesellschaft, offered the directorship to Hahn; Rust, as cognizant minister, filled the state-sup-

28. LS, *3*, 31; Haberditzl, in Harig and Mette, eds., *Naturwissenschaft-Tradition-Fortschritt* (1963), 324. Cf. ADB, 113–16; Hermann, *New physics* (1979), 84–85.

29. Text in MPG, *50 Jahre* (1961), 190, and in Hahn, in MPG, *Mitteilungen* (1960), 3–4.

ported position with a caretaker, Professor August Gerhart Jander from Greifswald, and gave him his Nazi student, Rudolf Mentzel, as his führer. Planck and Glum appealed, in the hope of reinstating Haber. Glum tried to interest the military in physical chemistry as a military science; Planck went directly to Hitler, with what results we already know; Jander and Mentzel remained; and, to make matters clear, Mentzel showed up at a meeting of the Society's Senate, in which he now represented the government, in an SS uniform complete with revolver.[30]

Haber died in Switzerland on 29 January 1934 of chronic heart disease. Laue published praises of his character, his intolerance of pettiness, and his devotion to country, and urged Planck to permit the Kaiser-Wilhelm-Gesellschaft to sponsor a memorial service on the first anniversary of Haber's death. After much internal debate, the Society's "protective dam," as Haber had called Planck, agreed to let its official tears flow.[31] Rust tried to stop the affair, on the ground that Haber had been an open critic of the new Germany. Planck refused to budge. Rust prohibited all state employees under his jurisdiction from attending, and other government offices followed suit. The service nonetheless took place with dignity and in a well-filled auditorium; it offered the attendees—foreign dignitaries, nongovernmental employees of the Society, wives of Rust's civil servants, representatives of the military and big business—the enjoyment of a successful, if transitory, expression of defiance.[32]

30. Hahn, in MPG, *Mitteilungen* (1960), 4; Glum, *Wissenschaft, Wirtschaft und Politik* (1964), 447–50; Planck, *PB, 3* (1947), 143; ADB, 57, 171.

31. Laue, *Nwn, 22* (1934), 97; Planck to Laue, 22 Mar. 1934, in AH, 86; Haber, *Nwn, 21* (1933), 293.

32. MPG, *50 Jahre* (1961), 190–94; Hahn, in MPG, *Mitteilungen* (1960), 6–13; Glum, *Wissenschaft, Wirtschaft und Politik* (1964), 440–41. Cf. ADB, 67–68; Hermann, *New physics* (1971), 86; Willstätter, *Aus meinem Leben* (1949), 272–77.

In 1934 and 1935 Planck could still have regarded his policies as fruitful, at least in the Academy. It had defeated Stark's candidacy and refused its formal approval of Stark's appointment, in June 1934, as Schmidt-Ott's replacement at the Deutsche Forschungsgemeinschaft (German Research Community), the successor of the Notgemeinschaft. And it was allowed to go its own way at public expense, perhaps because the Nazis regarded it as marginal to their science policy. It retained its Jewish members, other than Einstein, and its Jewish employees, whom Planck helped to protect. In the opinion of a well-placed observer, Planck's reputation shielded the institution. "In the Prussian Academy of Sciences there are still [despite the Einstein affair] some of the 'righteous,' for whose sake the Lord would have spared the Academy. . . . I name first of all Planck, whose purity no one who knows him doubts."[33]

The Kaiser-Wilhelm-Gesellschaft also retained many if not most of its Jewish members, both in its Senate and in its institutes. But whereas the Academy lost only Einstein, the larger and more important Kaiser-Wilhelm-Gesellschaft had to discharge all the "Jewish" members of Haber's institute and to tolerate the presence of Nazi agents throughout its organization. Even the departments of Otto Hahn and Lise Meitner included party members, and the young and ambitious considered whether joining the party might not be necessary to a successful career in the Society's laboratories. A good example is Paul Harteck, a physical chemist who took a leave from his assistantship in 1933 to work in the Cavendish laboratory in Cambridge. Rumors reached him during the disturbance caused by Haber's resignation that his colleagues were joining the party. In November Harteck learned that owing to "reorganization" he would not be able

33. LS, *3*, 47–49, 372; ADB, 117–18; Richter, *Forschungsförderung* (1972), 21; Arnold Berliner to Ehrenfest, 25 Apr. 1933, in Pauli, *Briefwechsel* (1979), *2*, 147 (quote).

to return to his assistantship. Should he become a Nazi? Should he call in his "idiot cousin," a big Nazi in Austria? Should he appeal to Stark, or to Planck? Harteck went to Berlin and met with Planck, Hahn, and Meitner; eventually, he got a place without joining the party.[34]

In this atmosphere the Society protected itself by as superficial an "alignment" (*Gleichschaltung*) as possible to the regime. It took Nazis into its Senate, flew the swastika, ended its correspondence with "Heil Hitler!," and offered the Hitler salute on official occasions. An eyewitness to one such affair, P. P. Ewald, remembered Planck's execution of the salute. "Planck stood on the rostrum and lifted his hand half high, and let it sink again. He did it a second time. Then finally the hand came up, and he said 'Heil Hitler.' "[35] Here again informed contemporaries thought that Planck's presence ameliorated the treatment the Society would otherwise have received.[36]

The Society protected itself by mobilizing two interests the Reich did not care to alienate: big domestic industry and international public opinion. Planck seldom missed an occasion to invoke the regime's concern to project a brilliant image in the world and to stress the Society's international relations. "Quite apart from the fact [he wrote in the Soci-

34. Krafft, *Fritz Strassmann* (1981), 43–44; Harteck to Karl Bonhoeffer, 17 and 28 June, 8, 15, and 24 Nov., 26 Dec. 1933 (Rensselaer Polytechnic Inst., Troy, New York). On a scale where 1 = anti-Nazi, 3 = Nazi, Harteck was rated 2 by British Intelligence (list in Chadwick Papers 4/56, Churchill College, Cambridge). Cf. Scherzer, in Flitner, ed., *Deutsches Geistesleben* (1965), 50–52, on the pressure brought by Nazi agents to coopt young scientists.

35. ADB, 42–43, 60–61; *Science, 77* (1933), 529; Glum, *Wissenschaft, Wirtschaft und Politik* (1964), 471; Allan Gregg, diary, 15 June 1934 (RF 717/2); ADB, 1; LS, *3*, 32.

36. *New York Times*, 12 Jan. 1936, 3:2; cf. Planck to R. B. Goldschmidt, 12 July 1936 (Goldschmidt Papers, TBL).

ety's report for 1934–35] that these relations are of great importance for German scientific research, with their help many erroneous views abroad about the new Germany can be set right."[37] With this he made explicit the missing term in an argument he had made the previous year, when he had called on the Reich to supply German science generously as "the most effective means to deflect baseless research prejudicial to German interests." In other words, government support of true research, as represented and advertised by the Society, would counter charges that the new regime's interest in science did not go beyond the "scientific" study of race. This and similar disingenuous arguments carried weight. Support increased, and not only for institutes allied with practice. The previously immaterial physics institute acquired substance. And Planck produced the trade-off: "With a full understanding of [the Society's] high purposes," he wrote in his account of its activities for 1935–36, "the National Socialist regime is granting the support that is so necessary to the rebuilding of the German nation."[38]

RESIGNATION AND DISMISSAL

Lenard and Stark anticipated that Planck would have to resign as president of the Kaiser-Wilhelm-Gesellschaft at the end of his first term of office. That would give them their chance, Stark crowed, "to make something sensible of this completely pure Jewish business [the Kaiser-Wilhelm-Gesellschaft], which, as a start, must simply be pulled to pieces." But Planck neither resigned nor aligned. He tendered "reverential greetings" to the chancellor at the Society's first general assembly after the Nazi takeover and then gave a speech on "Deutsche Physik" that the Aryan physi-

37. KWG, *Nwn*, *23* (1935), 411.
38. KWG, *Nwn*, *22* (1934), 339, and KWG, *Nwn*, *24* (1936), 19. Cf. Planck's broadcast for foreign consumption, in HH, 97–100 (1935).

cists recognized as a travesty, without "a whiff of the new German spirit."[39] When Planck's second presidential term expired on 31 March 1936, Stark and Lenard conspired to insure that neither he nor anyone like him remained in charge of the Society.

The move against Planck was part of the campaign for Deutsche Physik, as Lenard entitled his general physics text, or "really true" science, as Stark ennobled it. The great physicists, according to Stark, had all been Germans; and, as an example of Nazi faithfulness to reality, he named as carriers of the German spirit Galileo, Newton, Faraday, Rutherford, a great line of Teutons whose current representative was Lenard. Shortly before the war a competitor entered the lists—this is still the gospel according to Johannes Stark—an abstract, arbitrary, mathematical, dogmatic, artificial physics, which defied the intuition and misrepresented the phenomena, a Jewish physics of which Einstein was the ringleader and relativity the climax. After the war this antiphysics prospered with the support of the international Jewish conspiracy, which had been the only real winner in the slaughter.[40]

Among the institutions captured by the Jewish conspiracy was the Nobel prize, which formerly went to solid experimentalists like Lenard and Stark, but which recently had fallen to Planck, Einstein, Schrödinger, and Heisenberg. The awards to the last two, in 1933, were particularly offensive

39. Stark to Lenard, 3 May and 20 Aug. 1933, in Kleinert, *PB*, *36:2* (1980), 36; Planck, "Deutsche Physik," *Berliner Börsenzeitung*, 25 May 1933; "The scientific situation in Germany," *Science*, 77 (1933), 529.

40. Stark, "Der germanische Galilei," *Nationalsozialistische Erziehung* (1937), 105, in Poliakov and Wulf, eds., *Das Dritte Reich* (1959), 288–89; Lenard, *Deutsche Physik* (1936), *1*, "Vorwort"; Scherzer, in Flitner, ed., *Deutsches Geistesleben* (1965), 53; Brüche, *PB*, 2 (1946), 232–36; Lenard, "Erinnerungen" (AHQP), 130, 134. Cf. the declaration of Heidelberg science students, 1936, in Poliakov and Wulf, *Das Dritte Reich* (1959), 305.

to Aryan physicists. Both sides recognized the political importance of the award. Planck wrote Schrödinger that it would aid the cause of theoretical physics, "which in our current circumstances appears especially desirable and timely."[41] Stark tried to use his new power as head of the Forschungsgemeinschaft to force German Nobel prizewinners to declare their support of a referendum to name Hitler president of the Reich. Despite Stark's detestation of modern theoretical physics and its representatives, he wanted their signatures. They refused, on the ground that the declaration was political, hence not a matter of scientific judgment. On this ground there was, however, much room for nuance, as appears from the contrasting replies of Laue and Heisenberg, as preserved and answered by Stark.

Laue refused without qualification, adducing the immiscibility of science and politics. Stark replied that the declaration was not political. "In fact it was to be a part of the great national profession of faith by the German people in its leader Adolf Hitler." That was of course nonsense. But then he added, quite rightly, that Laue had made a political statement in his public support of Einstein, who, to Stark, was a "traitor of the country and a slanderer of the National Socialist regime." Heisenberg's answer, as given by Stark or by someone informed by him, appeared in the SS's paper, *Das Schwarze Korps*: "Although I personally vote 'Yes,' it seems to me that political declarations by scientists are wrong since they are most unusual." Accordingly he refused his signature. "This answer [the paper editorialized] shows the Jewish spirit of its author."[42]

41. Planck to Schrödinger, 10 Nov. 1933 (AQHP/84).
42. Stark to Laue, 21 Aug. 1934 (DM); *Das Schwarze Korps*, 15 July 1937, 6, in Poliakov and Wulf, eds., *Das Dritte Reich* (1959), 307–8; Stark, "Einstein, Heisenberg, und Schrödinger," *Nationalsozialistische Monatshefte*, no. 71 (Feb. 1936), 106–7, in ibid., 301; D. Hoffmann in Humboldt, Univ., *Wissenschaft* (1982), 99–100.

The political use of the Nobel prize appears further in the protection it afforded its dissident holders, like Laue. For this reason Planck joined Laue and Heisenberg to try to force a prize for Hahn and Meitner, whom he had earlier nominated in the ordinary way for an award in chemistry. Their effort succeeded no better than Planck's championing of Sommerfeld and Paschen, whom the Swedish Academy steadfastly refused to honor.[43] The Nazis themselves made clear their worry about the political value of the Nobel prize after Carl von Ossietzky, a pacifist rotting in a concentration camp, won the peace prize for 1936. Hitler responded by prohibiting German nationals from accepting Nobel prizes. In Stark's fervid imagination, Heisenberg was the Ossietzky of physics.[44]

To grind up the Ossietzkys and "white Jews" of physics, who traitorously taught relativity theory and quantum mechanics, Stark and Lenard turned to the biggest cogs in the party machine. In May 1934 Stark combined with Ernst Gehrcke in trying to stir up the Prussian Kultusminister with the news that Arnold Berliner, as editor of *Naturwissenschaften*, and Max Planck, as general leader of theoretical physics, still supported Einstein and restricted the freedom of other physicists to think as they pleased. In October Lenard tried to mobilize the Reichsminister for Propaganda against the leaders of the "Einstein-clique, von Laue, Hei-

43. Planck to Laue, 22 Dec. 1936 (DM), and AH, 90–91. Planck nominated Hahn and Meitner in chemistry in 1929, 1930, and 1933, and Paschen and Sommerfeld in physics in 1923, 1925, 1926, 1928, 1932, and 1933. Otherwise, after his own prize gave him the right of perpetual nomination, he put forward Einstein (1919, 1921, 1922), Bohr (1922), A. H. Compton (1927), Heisenberg and Schrödinger (1930), and Otto Stern (1931), all of whom won eventually. Data from Nobel Archives, Kungl. Vetenskapsakademien, Stockholm.

44. Stark in *Das Schwarze Korps*, 15 July 1937, in Poliakov and Wulf, *Das Dritte Reich* (1959), 301; ADB, 158; *Science*, 85 (1937), 171.

senberg, etc."[45] But the mobilization against Planck did not build until 1936.

That January, as a shot in a campaign against Heisenberg in the *Völkischer Beobachter*, Stark pointed to Planck as a nefarious influence on German science. In February the *Nationalsozialistische Monatshefte*, another party paper under the control of Stark's patron, Alfred Rosenberg, Reichskanzler der NSDAP (Reich Chancellor of the National Socialist German Workers Party), carried a bolder sally against Einstein's creatures in positions of influence: Planck at the Kaiser-Wilhelm-Gesellschaft, Laue at the Academy, and Heisenberg, "the spirit of Einstein's spirit," whom Sommerfeld wanted as his successor in Munich. Then Gehrcke declared in the news magazine of physics, the *Physikalische Zeitschrift*, that it was owing to the machinations of the Einstein clique that Planck had grown famous for the radiation formula, an "elementary mathematical accessory," he said, in comparison with the measurements, "the original, fundamental, physical discovery," that it reproduced.[46]

The first of April came and went, and Planck remained at the Society. Stark, who perhaps had expected to complete his crown with the presidency of the Kaiser-Wilhelm-

45. Stark and Gehrcke to Preussicher Minister für Kunst und Wissenschaft, 27 May 1934, excerpted in Berliner to Planck, 28 July 1934 (DM); Lenard to Reichsminister für Volksaufklärung und Propaganda, 8 Oct. 1934, in Grundemann, in Dresden, Technische Hochschule, *Wissenschaftliche Zeitschrift*, 16:5 (1967), 1626. For the attack on Heisenberg, see Haberditzl, in Harig and Mettle, eds., *Naturwissenschaft-Tradition-Fortschritt* (1963), quoting correspondence of Rosenberg (Heisenberg belongs in a concentration camp but cannot be sent because of possible repercussions abroad), and ADB, 141–44, 153–68.

46. Gehrcke, *Physikalische Zeitschrift*, 37 (1936), 439–40, referring to Lenard, *Deutsche Physik*, 1 (1936), 12; Haberditzl, in Harig and Mette, eds., *Naturwissenschaft-Tradition-Fortschritt* (1963), 324–25; ADB, 142–44; Hermann, *New physics* (1979), 89.

Gesellschaft, had not been asked to take it. He urged Lenard to step forward, and so did Rust; Lenard declined, pleading age, the wisdom of which, however, he was willing to share. He revealed to the responsible official in Rust's ministry, the sometime applied mathematician Theodor Vahlen, that the Society was not an agency for the promotion of science, but a machine for the social legitimation of Jews. "From the beginning it was . . . a Jewish monstrosity with the purpose, entirely unknown to the emperor and his advisers, of enabling Jews to buy themselves respectability and of bringing Jews and their friends and similar spirits into comfortable and influential positions as 'researchers.'" The place had to be taken apart, Lenard shrieked—private donors, Jews, democrats, Freemasons, Planck, Laue, Peter Debye (called by Planck to the Kaiser-Wilhelm Institute for Physics), the whole bunch.[47]

In Lenard's dimming mind Planck's behavior needed special explanation. How could so good a German, so pure an Aryan, fail to use his large influence in the interests of the new Germany? Lenard wrote Planck to straighten his thinking about the nature of science, but, he complained, Planck did not profit from the instruction. "Furthermore, he showed himself so ignorant about race that he took Einstein to be a real German, who should be honored in his country." Lenard blamed this blindness on the many theologians and pastors in Planck's family. "The respect of these people for the Old Testament and their false interpretation of the life and death of Christ were and are a source of many errors, even the worst, which are now at work in the great war of the Jews against us."[48]

47. Lenard to Vahlen, 6 Apr. 1936, and Stark to Lenard, same date, in Kleinert, *PB*, *36*:2 (1980), 37. Some of these sentiments were published in *Das Schwarze Korps* in the summer of 1937, after Planck had retired; Schlicker, *JW*, 1975:2, 184; AH, 92–93.

48. Lenard, "Erinnerungen" (AHQP), 136–37.

As for Stark, who explained Planck as a creature of Einstein's and an addict of internationalism, his incapacity for much beyond plotting and invective soon neutralized his influence. Had he been less crazy, he would have been much more dangerous. His dictatorial manner, administrative incompetence, and identification with Rosenberg alienated Rust's ministry and particularly Mentzel, whom Stark had tried to bully. In November 1936 they forced Stark to resign as president of the Forschungsgemeinschaft, which soon fell from a policy-making to an administrative agency.[49] Stark still retained his post at the Reichsanstalt, however, and that soon posed a problem for Planck.

The golden anniversary of the Reichsanstalt fell in November 1937. Should Planck attend? His presence would dignify an affair presided over by a man who had openly striven to injure him, and it would imply continuity between the great imperial institution and the Nazi nest that carried on its name. Laue, Hahn, and Meitner urged him not to go. Planck weighed all—and went. "I keep coming back to the conclusion that the Reichsanstalt is more important than the person of Mr. Stark." The institution took priority over the person, continuity and appearance over transient reality. Moreover, there was a matter of gratitude, and even piety, toward the laboratory that did the experiments that made possible Planck's greatest work and toward the Siemens and the Helmholtz families that had created and inspired the original institution. "Against all this, Stark does not count at all."[50]

Planck continued at the Kaiser-Wilhelm-Gesellschaft for almost half of a regular presidential term while its Senate sought a suitable successor. The choice was I. G. Farben's

49. Richter, *Forschungsförderung* (1972), 58; ADB, 118–22; W. E. Tisdale to W. Weaver, 1 Aug. 1934 (RF 717/2), reporting the opinion that Stark was losing influence.

50. Planck to Laue, 17 Nov. 1937 (DM), partially in AH, 94.

Carl Bosch, an industrial scientist with a Nobel prize in chemistry, a former collaborator of Haber's, no Nazi but also no follower of Einstein. Rust, who required a reorganization that introduced the führer principle (according to which the head of an institution dictated to his subordinates), eliminated the powerful executive secretary Glum, and changed the makeup of the Senate. Although it never did align fully with the government, the Society's independence, which had already been greatly impaired, diminished further after Planck and Glum left in 1937.[51] In the renewed and harder persecution of the Jews that culminated on "crystal night" on the twentieth anniversary of the Armistice, the Society could no longer protect its remaining Jewish employees. Already the previous March, when the annexation of Austria took place, it had had to apply the German racial laws to Austrian citizens, notably Lise Meitner, in its institutes.[52]

It was not possible for the Academy to withstand pressures that refashioned the Kaiser-Wilhelm-Gesellschaft and ruined the universities. In October 1938 Rust required the Academy to introduce the führer principle, expel its three Jewish ordinary members, and accept limitations on its cooptation of members. It fell to Planck to request the resignations of his Jewish colleagues. They went quietly, as had the Jews refused admission to the University's physics colloquium the year before; and Planck doubtless thought of the former, as he said of the latter, that it was well that their undesirable exclusion occurred "without scandal."[53] The time for noise or protest had passed. By backstage diplomacy

51. Glum, *Wissenschaft, Wirtschaft und Politik* (1964), 487–90; KWG, *Nwn*, *36* (1938), 321; MPG, *50 Jahre* (1961), 195–96; ADB, 61. Planck supported Bosch's election; LS, *2*, 198; Marga Planck to Hans Hartmann, 29 Dec. 1946 (MPG).
52. Krafft, *Fritz Strassmann* (1981), 170–73.
53. Planck to Laue, 17 Nov. 1937 (DM).

Planck helped to save some shreds of the Academy's pride and independence before his own dismissal.

On 22 December 1938 Planck resigned his secretaryship after twenty-six years of service, to make way for the Academy's führer. Rust proposed to raise his lieutenant Vahlen, then only just elected an ordinary member, to dictator of the Academy. Planck led a fight to reject Vahlen, who had the double demerit of being a Nazi and an applied scientist, and to preserve the Academy's right to choose new members. Both battles ended in moral victory in June 1939, when Rust conceded a nominal right of nomination and had to break a tie vote between Vahlen and Planck's candidate for president, the geologist Hans Stille, who had briefly held the secretaryship resigned by Ficker in 1937.[54] Apparently Planck had had recourse to his favorite argument about the international image of German science. One of the fascist members of the Academy wrote Vahlen during the fray: "If Mr. Planck thinks that only eighty-year-old Nobel prizewinners are able to represent German science abroad worthily, let him not forget that you now look after the interests of German science at home, as far as they concern our Academy."[55]

Planck's retirement did not free him from the attacks of the Aryan physicists. He was hounded by the *Zeitschrift für die gesamte Naturwissenschaft*, a monthly yelping of the Reichsstudentenführung (Reich Student Command), which protected young people from "Jewish" influence. The Stu-

54. See Vahlen's inaugural speech and Planck's reply, 30 June 1938, in *Sb* (1938), xcviii–c; LS, *3*, 63–68, 72, 157–58, 166, 201. The elections in physics 1936–39 brought only first-rate men: Debye, Hans Geiger, Ludwig Prandtl, Heisenberg (in 1938, despite the steady attacks on him in 1937–38), and Walther Bothe. Ibid., 177–78.

55. Letter of 15 Mar. 1939, in LS, *3*, 70. Planck had juxtaposed the declining status of German science in the world with the situation and responsibility of the Academy in *Sb* (1937), liii.

dent Command barked at Planck mainly through the mouth of Hugo Dingler, a man esteemed by the Aryan physicists as one of their pioneers and by himself as a thinker of unparalleled clarity. Dingler held Planck to be the "founder and ringleader of the *unanschaulich*, or increasingly mathematical direction in physics." He declared that people misled by Planck misunderstood experimental physics and had no patience for thorough analysis in the Dinglerian manner; they spun world pictures from a few isolated threads, ascended immediately to mathematics, "which rules out any real thinking," and proclaimed their mathematico-physical novelties as fundamental truths. These faults characterized Jewish thinking, as explained by Lenard and also by Stark, who had recently enlightened the British in the matter.[56] It followed, so Dingler declared in lectures as well as essays, that Planck at best was no physicist, and that his famous formula, as Gehrcke had already made clear, was but a "mathematical reworking stumbling after experimental results."[57]

No doubt Dingler—and also Stark's old porcelain student Glaser and Sommerfeld's unworthy successor Wilhelm Müller, who supported Dingler's opinions—were ridiculous.[58] They were also dangerous. Planck may have laughed at their prediction that modern physics would not survive their attack. But his frequent lecturing around 1940 had as one objective to drown out the harangues of the Aryan physicists by the constant quiet repetition of his own beliefs.[59]

56. Stark, *Nature, 141* (30 Apr. 1938), 770–72; Lenard, in *Zeitschrift für die gesamte Naturwissenschaft*, 5 (1939), 41.

57. Dingler, *Zeitschrift für die gesamte Naturwissenschaft*, 4 (1939), 329, 339–41, and ibid. 5 (1939), 47, 57; cf. Dingler, *Methode der Physik* (1938), 396. Dingler put his several attacks together as *Max Planck und die Begründung der sogenannten modernen theoretischen Physik* (1939).

58. Wilhelm Müller, *Zeitschrift für die gesamte Naturwissenschaft*, 5 (1939), 173–74, and 6 (1940), 283; Glaser, ibid., 5 (1939), 275.

59. Dingler, *Zeitschrift für die gesamte Naturwissenschaft*, 4 (1939), 341; Planck to Artur Neuberg, 24 Apr. 1940 (ANP); Neuberg to Planck,

THE ARK

Planck's achievements as science administrator included more than help to Jewish colleagues and employees in the temporary conservation of their status and income. It was his personal triumph that a Kaiser-Wilhelm Institute for Physics came into existence in 1938. As we know, a physics institute had been a high priority of the Kaiser-Wilhelm-Gesellschaft before the First World War, and was realized in 1917 only as a mechanism for dispensing grants for ongoing research projects. In the spring of 1929 Planck and the other major Berlin physicists moved to establish a Kaiser-Wilhelm institute for theoretical physics. The timing seemed right both financially and substantively: the Republic was prosperous, and theoretical physics had recently undergone a deeper and more consequential development than it had ever experienced before. The proposed institute would have experimental facilities, devoted first of all to X rays and molecular beams. Without the means of quick confirmation or refutation, Planck and his colleagues argued, theorists would lose direction and time; they therefore required a four-story building with the latest apparatus and the best experimenters.[60]

Just after the financial crash of 1929, the projectors, to be on the safe side, submitted the same proposal to the Academy. It was not a good time to ask the Republic for money. In December the Kaiser-Wilhelm-Gesellschaft exported the proposal to the Rockefeller Foundation, which had helped with the financing of other institutes. In a few months the Foundation granted what German science, industry, and

---

17 May 1938 and 25 Jan. 1939, re a lecture by Dingler, and 11 May 1942, re Planck's success (ANP); Planck to Neuberg, 13 Nov. 1941 (Autog. I/ 443, SPK).

60. Planck to Laue, 24 Nov. 1929 (DM); Einstein et al. to Harnack, 5 May 1929, in Kirsten and Treder, *1*, 154–57.

government had declined to provide in a decade and a half: the contemporary dollar equivalent of 1.5 million RM for building and equipping a physics institute to be headed by Laue, and an additional sum to acquire the necessary land in Berlin-Dahlem. There was the condition that, as usual with the Foundation's grants, the recipients promise to assume the cost of operations.[61]

Glum accepted the terms but doubted the possibility of compliance, since the Kaiser-Wilhelm-Gesellschaft had for maintenance only the income (about 50,000 RM) from the sum it had reserved for purchasing land, and the hard-pressed government refused to promise the additional 100,000 RM thought necessary. A request from Laue that the Society be allowed to build part of the institute and apply the income from the rest of the grant to operations was not favorably received by the Rockefeller Foundation. In 1931 Glum and Planck asked to postpone building for a year or two; meanwhile the money might be found, and Planck would be able to establish physics in Berlin on a sound basis. He had in mind obtaining additional support from the government by coupling the directorship of the hoped-for institute with the professorship of physics at the University of Berlin that would soon be released by Nernst's retirement.[62] The combination might attract not only state money but also (and more likely) the best available physicist. Planck hoped to entice Franck from Göttingen and to transform the paper theoretical institute into a brick-and-mortar home for experiments. He spared no effort to make the position as attractive

61. L. W. Jones, diary entries, Feb. 1930, and to RF, 5 Mar. 1930; RF to KWG, 14 Apr. 1930, and Gregg and Jones to Harnack, 1 May 1930 (RF 717/1).

62. Glum to Gregg and Jones, 1 May 1930; "H. A. S.," diary entry, 15 Dec. 1930; Jones, diary entries, 2 Jan., 30 Mar., 12 July 1931, 7 Apr. 1932; Glum to Gregg, 1 Oct. 1931, all in RF 717; "Niederschrift über die gemeinsame Sitzung des Kuratoriums und des Direktoriums des KWI für Physik," 14 July 1931, in Kirsten and Treder, *I*, 158.

as possible, as he told Laue, who had effectively taken over Einstein's directorship but agreed to step down to second in command. The plan did not work: the budget of the Kaiser-Wilhelm-Gesellschaft had to be cut again, Nernst stayed another year at the University, the Nazis came, Franck emigrated, and the institute for physics stood poised to return to nothingness.[63]

As a Rockefeller representative said, "The completion of the Physics Institute lies very close to Planck's heart."[64] It had become for him the ark in which a few exemplary German physicists could weather the national storm. By June 1934 he had picked his Noah in the person of Peter Debye, head of experimental physics at the University of Leipzig, where he worked harmoniously with Heisenberg. Besides his scientific attainments, Debye had the advantage of Dutch citizenship and the toughness to stand up to Nazi bureaucrats; he justified choosing his assistants for their scientific rather than for their political merit by citing the führer principle, which, he said, made him dictator in his laboratory. Debye's lieutenant would be Laue, who would oversee x-ray research in the new institute. But to obtain release of the Rockefeller grant, Planck still required 100,000 RM a year. In July 1934 he wrote the Foundation that the government had agreed to provide the sum.[65]

The Rockefeller field representatives doubted that the agreement meant much. For one thing, it was only verbal. For another, they thought it likely that the government would fall and, if it continued, that it would repudiate any

63. Planck to Laue, 10 Nov. 1931 (DM), and to Einstein, 4 Mar. 1933 (AE), still hoping for Franck; Einstein to Planck, 15 Nov. 1930 (AE), and to Laue, 12 July 1922 (MPG A/39), on Laue's de facto directorship. Cf. KWG, *Nwn, 20* (1932), 428.

64. W. E. Tisdale to Warren Weaver, 4 Sept. 1934 (RF 717/2).

65. Tisdale, diary entry, 12 June 1934; Planck to Max Mason, 6 July, and to RF, 3 July 1934 (RF). Like Debye, Planck pretended that the KWG followed the führer principle by giving directors full authority in their laboratories. Planck in HH, 91 (1935).

commitment to pure science. Already the institute for chemistry worked largely on chemical warfare (so the men in the field reported), and an institute for anthropology devoted itself to race purification. Planck, "who has held the KWG on a straight course," was old, and Stark stood ready to take his place. The New York office answered that it knew "comparatively little about [Stark] as a person," thereby declaring its incompetence to decide on the release of the funds; it worried instead about covering the running expenses.[66] Meanwhile, on 26 July 1934 the Foundation's W. E. Tisdale visited Planck and asked whether a written commitment could be secured. "He threw up his hands [Tisdale reported] and said that negotiation with the Government was almost impossible. . . . Negotiations are interminably slow and met at every step by indecision and red tape."[67]

Tisdale's visit prompted Planck to make a personal appeal. For four years, he said, he had worked hard to bring the institute into existence; were the Rockefeller Foundation to withdraw now, "it would be an extremely painful disappointment for me." It would sink the ark. "The future development of physical science in Germany depends very much on whether we are now finally successful in creating a modern physical institute of the first rank, which we have sadly missed for some years in Germany."[68] This appeal by the conscience of German science helped the Foundation to see its way, against its inclinations, to honor the commitment it had made five years before in a different political environment. So perhaps did an interview with Franck, who said that the Nazis' depreciation of pure science "would cause him grave doubts" about the wisdom of building the institute were it not for the presence and guarantees of Planck. On 1 November 1934 the Foundation decided to release the

66. T. B. Appleget to Mason, 30 July 1934; Tisdale to Weaver, 4 and 24 Sept. 1934, and Weaver to Tisdale, 14 Sept. 1934 (RF 717/2).
67. Tisdale to Weaver, 1 Aug. 1934 (RF).
68. Planck to Tisdale, 29 Aug. 1934 (RF).

funds and requested a written commitment from the government, which, after much negotiation, Planck was able to supply the following February. In a further coup, he persuaded the authorities to credit the Kaiser-Wilhelm-Gesellschaft with 1.5 million RM for the Rockefeller grant, although at the official rate of exchange the dollars would have brought less than a million marks.[69]

Then Debye went into action. He got the government to agree to give double its promised contribution to operating expenses within two years and to allow him to name his assistants without restriction; that made Debye "the only undepressed person" Tisdale found in Germany in the fall of 1935. Scientific work began in the institute in 1937. It was not dedicated, however, until 30 May 1938. The reason for the delay, Debye told a visitor from the Rockefeller Foundation, was that Stark and Lenard had protested against the name Debye had chosen for the institute and had mobilized the Ministry for Science, Education, and Popular Culture. This conversation took place in January 1938. Five months later Debye felt himself strong enough, and Stark and Lenard sufficiently discredited, to open his new research facilities, with great fanfare, as the Max-Planck-Institut für Physik.[70]

As Chaplain

In the later 1930s Planck protested against the worldview of the government in the only way his temperament and outer circumstances permitted: reprimands or counter-affirma-

69. Tisdale, diary entry, 28 Oct. 1934 (quote), Mason to Planck, 1 Nov. 1934, and Planck to Mason, 12 Feb. 1935 (RF); Planck to Debye, 11 Feb. and 11 Apr. 1935 (MPG A/39).

70. Tisdale, diary entry, 4 Oct. 1935, Weaver, memo to file, 21 Jan. 1938, and Debye to Weaver, 31 May 1938 (RF 717); "Aktennotiz" of meeting of 26 May 1937 at the Reichsministerium für Wissenschaft, Erziehung und Volksbildung (MPG A/39). The Society had long since accepted the name; Debye to Ernst Telschow, 26 Feb. 1936 (ibid.).

tions inserted in speeches dealing with other matters. An early example is his *Physik im Kampf um die Weltanschauung* (*Physics struggling for a worldview*), given at Harnack-Haus, the Kaiser-Wilhelm-Gesellschaft's cultural center, in March 1935.

The argument is familiar: the basis of science does not lie in the nature of things; "a certain dose of arbitrariness" must be admitted at the beginning; it proves most useful and productive to stipulate a causal, real outer world; causality extends to history and psychology, which, indeed, must presuppose determinism to have any claim as sciences; no science can define good behavior, and to act well we must be guided by our feeling of personal freedom and disciplined by a religious sensitivity. The argument ends in a new place, however. Although ethics, like science, springs from national roots, neither deserves its name unless it transcends its origins; both aim at a universal truthfulness and objective justice, equal justice, for all. "Woe to a commonwealth if its feeling of the security of law falters, if consideration of place or origin plays a role in judicial proceedings." By following the highest ideals of truth and justice, Prussia and the Reich became great: "May they never be lost to our people."[71] A similar, and similarly veiled, criticism of the regime occurs in Planck's Leibniz Day speeches for 1935 and 1937. In the first he observed that Leibniz had been able to combine deep national feeling with cosmopolitanism, open-mindedness, and equal-handedness. In the second he read several lessons from the history of the Academy, notably its decline after Germanification at the end of the eighteenth century ousted its leading member, the cosmopolitan mathematician Joseph-Louis Lagrange.[72]

Late in 1936, before the Leipzig section of the German

71. *Wege*, 259–60, 263–65, 270–71, 272 (quote).
72. Planck, *Sb* (1935), cii–ciii; *Sb* (1937), xxi–xxii. Cf. HH, 58–66.

Philosophical Society, Planck illustrated his standard resolution of the problem of the freedom of the will with bold and even startling references to contemporary circumstances. He adduced as a model of an ethical dilemma the situation of a man who, having hidden a friend he knows to be innocent of any crime, is questioned by the police. He has a choice, or rather, he feels he does, between lying and betrayal. To pick the lesser evil he needs a firm ethical guide, which he will not find in the popular doctrines that deny the value of the individual or in any "immature social theory that has destroyed impartiality and set aside natural inhibitions."[73]

Honoring colleagues provided Planck with several occasions for more direct criticism of Nazi policies. In June 1937 the Physical Society celebrated the silver anniversary of the discovery of x-ray diffraction by Laue, Walter Friedrich, and Paul Knipping. Planck praised their work as a splendid example of the fertility of collaboration between theorists and experimentalists, and he disparaged for their ignorance people (meaning Stark and his followers) who believed that the problems of atomic physics could be resolved by returning to the "intuitions of the classical era." Contemporary theory no doubt clashed with human intuitions, but that lay in the nature of things and signified that "theoretical physics has never been closer to reality than it is at present."[74] Later in the year, in a tribute to Sommerfeld, Planck explicitly alluded to the plan of the Aryan physicists to install an incompetent party member rather than Heisenberg as Sommerfeld's successor in Munich. "May the quiet satisfaction be allowed him [Sommerfeld] to see one of his worthy disci-

73. *Wege*, 277–78, 280 (quote), 286–87. Cf. the review by Bernhard Merten, *Nwn*, 25 (1937), 541–42.

74. *PA*, *3*, 366–67 (1937); LS, *3*, 61. Cf. Planck, *Nwn*, 27 (1939), 666, praising Laue for keeping alive theoretical physics at the University of Berlin.

ples, who is scientifically his equal, in his place."[75] And the following year, Planck climaxed the grand celebration of his own eightieth birthday, arranged by the unaligned Physical Society, by awarding the Planck medal to Louis de Broglie, who had the twin blemishes, to the Nazi ideologues, of being a theorist and a Frenchman.

Planck had chosen de Broglie for reasons not entirely scientific. Or so the matter appeared to Sommerfeld, who, as a previous winner, had by statute the right of approval of nominees. Sommerfeld preferred Enrico Fermi as the better physicist. "But I could not help but honor Planck's request without contradiction. Politically the order (1) de Broglie (2) Fermi is indeed more congenial, but thankfully in this case that plays no part."[76] On the contrary: had Fermi been chosen, Planck would not have been able to address the prizewinner as a citizen of a nonfascist state, a neighbor wishing to live in peace, a neighbor with whose fate Germany's was intertwined. "May a good fate grant that France and Germany come together before it is too late for Europe." To which the French ambassador, who accepted the award, replied: "In Geheimrat Planck we know and hail not only one of the brilliant founders of modern physics but also a complete type of man, of whom his country and the whole world have a right to be proud."[77] Planck would have preferred to hand the medal to de Broglie, but—and this is a measure of his mixing of science and politics—he considered the ambassador, rather than a French scholar, the best substitute.[78]

75. Planck, *Nwn*, *26* (1938), 779.

76. Sommerfeld to Einstein, 30 Dec. 1937, in Einstein and Sommerfeld, *Briefwechsel* (1968), 118; Richter, *Forschungsförderung* (1972), 58. Sommerfeld had come to see politics from Planck's point of view: Planck to Laue, 22 Oct. 1941 (DM).

77. *PA*, *3*, 411–12 (1938). Cf. AH, 93–97.

78. Planck to Laue, 10 Apr. 1938 (DM).

A final and subtler example. At the same eightieth-birthday celebration, Planck singled out for mention among his students only Max von Laue and Moritz Schlick. The reasons for distinguishing Laue—his closest, highest-ranking, and most courageous disciple—are not far to seek. But why the Austrian Schlick—Schlick who had forsaken physics for philosophy, and for a philosophy anathema to Planck? Schlick was a positivist, eventually successor to Mach in Vienna, and an outspoken critic of Planck's conception of science. But he was also an outspoken critic of Nazi ideas: it is likely that Planck recalled Schlick's memory (Schlick had been killed by a crazy student in 1936) as an indication of his own political views and his opposition to the German takeover of Austria.[79]

Just before his eightieth birthday, when he might reasonably have thought to retire from public speaking, Planck entered into a new career. He became an itinerant preacher. The role, of course, was in the family. He had been raised an observant Lutheran and regularly won prizes at the gymnasium in religion and deportment; he said prayers at table; he never doubted the value of organized religion and even as a young man was shocked at his friend Carl Runge's skepticism. From 1920 until his death in 1947 he served as an elder (*Kirchenältester*) in the congregation in Berlin-Grünewald.[80] During the Weimar years he had referred occasionally in general lectures to the easy reconcilability of science and religion; and in a newspaper article published on Christ-

79. *PA, 3*, 415; Herneck, *Wissenschaftsgeschichte* (1984), 79–81, 85–86. Both Planck and Einstein had a high regard for Schlick's ability: Planck to Wien, 13 June 1922 (AHQP); Einstein to Born, 9 Dec. 1919, in Einstein and Born, *Briefwechsel* (1969), 38–39.

80. AH, 9, 13; Iris Runge, *Carl Runge* (1949), 24–26; Herneck, *Wissenschaftsgeschichte* (1984), 89; Planck to Frieda and Hans Clarke, 17 May 1936 (APS), mentioning lectures in Vienna, Budapest, Graz, Basel, Bern, Geneva, Zurich; Planck to Laue, 3 Mar. 1944 (DM), mentioning table prayers.

184 · The Dilemmas of an Upright Man

mas Day 1930, he construed in favor of religion his usual argument that doing science requires belief in something extrascientific. Had people understood this requirement, he said, they would not have deserted the Church for false prophets, who proclaimed a suppositionless science or a carefree hedonism.[81] But it was not until May 1937 that Planck opened his mind fully on the topic, in a beautiful lecture entitled *Religion und Wissenschaft* (*Religion and science*) that distilled his thought about the nature and unity of knowledge.

The lecture grows around the question whether a man of science can be truly religious. By religion Planck meant the "binding of man to God"; religion, common to all men, rests on respect and awe before a superior power. What has set people at one another's throats is not religion but its trappings and arbitrary symbols, which are but the crudest means of grasping this power, or God. The religious man cannot prove logically the objective existence of God; he believes, as the condition of his religiosity, that God exists independent of earth and man and that from eternity He has held the world in His all-powerful hand. The creative man of science, too, has a preconditioning belief in an independent outer world, in "a universal order that we can know to some degree." Here the true nature of Planck's bête noire, positivism, is revealed: it bears the same relation to science that agnosticism does to religion. And there is yet another parallel. Like the religious man, the true scientist realizes that he can never understand fully the object of his preoccupation. He wonders at the ability of his trivially small mind to discover the elementary building blocks (the universal constants!) and laws of nature, and to recognize in the widest of them, the principle of least action, the presence of

81. HH, 75–80.

a teleological element, the demonstration of a "rational world order."[82]

The next step a child could take: our drive toward unity obliges us to identify "the world order of science with the God of religion." There is, however, this difference: for the religious man, God stands at the beginning, for the scientist, at the end, of all thinking. We must believe to act ethically, and we must act; society could not survive if its members went about without proven moral precepts or waited until acquiring wisdom to decide how to behave. Therefore each individual must strive to develop both sides of his nature, the religious and the scientific, which complete and complement one another: "It is the steady, ongoing, never-slackening fight against scepticism and dogmatism, against unbelief and superstition, which religion and science wage together. The directing watchword in this struggle runs from the remotest past to the distant future: 'On to God!' "[83]

Religion and science, given first to audiences in the Baltic provinces, proved astonishingly popular. It was printed five times in less than two years, abstracted and excerpted in many newspapers, and advertised by the Evangelischer Pressverband (Protestant Press Union), which considered it to mark the end of a century of warfare between science and belief. We are told that in that bottomless well of anti-German sentiment, Catholic Belgium, Planck's essay was widely approved, although it did not ascend all the way to a personal God.[84] Whence the interest? Ficker remembered that when Planck lectured in Vienna the final line, "On to God!" called

82. *Wege*, 293–94, 297, 304 (1937).

83. *Wege*, 304–5, 306. Cf. Kropp, *Zeitschrift für philosophische Forschung*, 6 (1952), 450–53.

84. *Das evangelische Deutschland*, *14:23* (6 June 1937), 197; Planck to Frieda Clarke, 17 May 1937 (APS); Daisomont, *Max Planck* (1948), 6, 12.

forth wild applause. The quiet optimism and calm delivery of the lecturer, his synthesis of pure science with universalistic religion, his union of rationality and belief, made an impression, and perhaps encouraged a hope, in those last years before Armageddon.

Optimism, affirmation of life, was Planck's message and the core, he said, of every wholesome philosophy. In 1934, inspired by Albert Schweitzer's book on Indian thinkers, Planck had written that the problem of the denial or affirmation of life was among the most important features of our worldview. In 1938 he returned to the theme, having learned that Schweitzer esteemed Chinese sages over Indian ones. That suited Planck, since the Chinese, in contrast with the Indians, built their philosophy on affirmation. "In my opinion every philosophy has the task of developing an understanding of the meaning of life, and in setting up this task one supposes that life really has a meaning. Therefore whoever denies the meaning of life at the same time denies the precondition of every ethics and of every philosophy that penetrates to fundamentals."[85]

During the war Planck's speeches took a somberer tone. His last major talk, *Sinn und Grenzen in der exakten Wissenschaft* (*Sense and limits of exact science*), first given at Harnack-Haus in November 1941, ends in resignation. To be sure, we find much of the old optimism and its grounds: the easy escape from positivism and its evils; the literally wonderful progress of science; the healthy belief, which underlies all science, in the postulate of an unattainable real world; the symbiosis of pure and applied science, illustrated ominously by Hahn's discovery of nuclear fission and the prospect of a cheap energy source. But the optimism is qual-

85. Planck to Schweitzer, 16 Nov. 1934 and 9 Aug. 1938, in Kangro, *Vorgeschichte* (1970), 236.

ified: the uranium machine might destroy the planet, and exact science gives no reason to think that things will get better. We are not born with a right to happiness. "And so we must consider every kindly gift of fate, every hour contentedly passed, as an undeserved gift and an obligation."[86]

Science alone could not console, nor even science and religion. There remained only resignation. Planck now counseled "a courageous perseverance in the struggle of life and a quiet submission to the will of the higher power" and a clinging to the only possession that no worldly power could steal, "a clear conscience, which expresses itself in scrupulous fulfillment of duty."[87] Underlying this sober advice, however, was a conviction and a message that things would get better, that some clear German consciences would survive into a decent world. Planck's attitude in the middle war years appears from his answer to a man he deeply respected, the great Goethe scholar Anton Kippenberg, who had written that it would be a blessing if a uranium bomb destroyed mankind. Planck replied that, when despondent, Kippenberg might find it more heartening to do as Planck did and look at things "in as scientific a manner as possible," rather than to contemplate the end of the human race. That meant to take the viewpoint of "an unprejudiced, objective observer. . . , who follows the course of events eagerly and with great interest, but who does not himself suffer from them." So fortified, Planck said, he had a burning desire to

86. *Wege*, 325, 331–34, 337–38, 339 (quote).
87. *Wege*, 339 (1941). Planck had used similar language before, but to a different purpose, for example, in ibid., 273 (1935), where "a clear conscience and a good will" are the fixed points of a progressive weltanschauung, and in ibid., 322 (1937), where one is advised to do one's duty, not because there is nothing else to do, but because it might confer "a blessing and prosperity on our dear German nation." Cf. Planck's broadcast of 1935, in HH, 104.

see the crisis through, to live long enough to experience the turning point, "the beginning of an improvement."[88]

Planck himself brought his tidings all over Germany and as far as Zagreb and Rome. The captive press distributed them in Poland, Norway, the Netherlands, and doubtless elsewhere, and Planck was allowed visits to neutral Switzerland and Sweden.[89] Apparently some influential parts of the Nazi government and the military considered that there was propaganda value in his preaching. And in his person. A unit of the propaganda ministry made a film of Planck in 1942 for its Archive of Celebrities (Filmarchiv der Persönlichkeiten); apparently the propagandists imagined that clips of the famous and respected, speaking informally about their life and work, might be useful in cultural warfare.[90]

Hans Hartmann, the nephew who had interviewed Planck over the radio in the mid-1930s and who was identified closely enough with the regime to be brought before a denazification board after the war, gave an account of his uncle's life and opinions in 1938 that recommended Planck, "Der Deutsche," in the cant and codewords approved by the

88. Planck to Anton Kippenberg, 17 May 1942 and n.d., and Kippenberg to Planck, 19 Mar. 1945. Planck and Kippenberg had become close in the Goethe-Gesellschaft before the war; Planck to Kippenberg, 2 May 1938, 27 May 1939, 10 July 1941 (KA). Similar upbeat messages occur in Planck to Laue, 10 Feb. and 19 May 1941 (DM), to Hildegard Gravemann, 20 May 1942 (MPG), and to Artur Neuberg, 3 Apr. 1942 (ANP).

89. Planck to Hartmann, 28 Apr. 1942 (visits to Dresden, Mülheim, Essen, Münster) (MPG); to Max Schirmer, 30 Sept. 1943 (visit to Kassel) (MPG); to Laue, 6 and 20 Sept. 1942 (visit to Zagreb, arranged by the German "representative" there), and 5 June 1943 (visit to Italy) (DM); to Artur Neuberg, 4 Dec. 1942 (visit to Budapest) (ANP); to Hildegard Gravemann, 18 June 1943 (visit to Sweden) (MPG); *Deutsche allgemeine Zeitung, 81* (27 Sept. 1942), front page; *Moselland kulturpolitische Blätter* (Apr.–June 1943), 69–71; Hale, *The captive press* (1934), 279–83; Martens, *Zum Beispiel "Das Reich"* (1972), 43, 258.

90. Kirsten, in AW, *Aus der Arbeit, 8:4* (1983), 12–13.

Reich.[91] At the same time, Erich Rudolf Jaensch, an expert on the race and its perfections, gave Planck as an example of a scientist of the "ideal German type."[92] The experts on the big lie, however, recognized that Planck did not advance the cause. An agency under Rosenberg, the Hauptamt Wissenschaft (Head Office for Science) of the NS-Kulturgemeinde (National Socialist Community), which kept dossiers on sixty thousand "scientists," paid unusual attention to Planck and judged that his universalism and refusal to repudiate relativity threatened "a great devaluation of the scientific worldview."[93] Although Deutsche Physik had been discredited and relativity rehabilitated,[94] invoking the name of its creator, like conjuring with spirits, could still be dangerous. Planck had observed the prohibition against mentioning Einstein's name in his major addresses during the late 1930s, although he had frequently discussed relativity, a compromise for which his reputation suffered abroad.[95]

A more disagreeable way of separating the creator from his creation occurs in *Das Weltbild der Physik* (*The world*

91. Planck to Hartmann, 18 Jan. 1946, re denazification (MPG). Planck probably agreed to the biography to help Hartmann's literary career; cf. Planck to Hans Hartmann, 19 Oct. 1932, 19 Oct. 1935, and 1 Jan. 1936 (MPG).

92. Erich Rudolf Jaensch, *Der Gegentypus* (Leipzig, 1938), quoted in Lindner, in Mehrtens and Richter, eds., *Naturwissenschaft, Technik und NS-Ideologie* (1980), 97. Cf. *PhW*, 226, which shrewdly compares later editions of Hartmann, which mention Planck's antifascist attitude, with the edition of 1938. The Plancks provided some biographical information, for example, in Planck to Hartmann, [May 1942] (MPG); "HH" is the fourth version.

93. Baumgärtner, *Weltanschauungskampf* (1977), 92.

94. On the rout of Aryan physics in 1940, see Benz, *Sommerfeld* (1975), 177–78; LS, *3*, 132–33, 140; and ADB, 177–79.

95. *Science, 84* (16 Oct. 1936), 35; Scherzer, in Flitner, ed., *Deutsches Geistesleben* (1965), 55.

picture of physics) by Planck's friend the Reverend Artur Neuberg, who was esteemed an honorable man. Although he pointed out the absurdity of judging scientific work from a racist standpoint, he nonetheless felt obliged (this in 1939) to deplore that relativity theory was the work of a Jew. A physicist who protected himself in anonymity wrote to Neuberg to criticize the remark about the authorship of relativity. Neuberg accepted the criticism and sent a copy of it to Planck, with this compromising explanation: the remark was "an unnecessary tribute to the spirit of the times, understandable because of my concern for [the success of] my book, but not suitable in the long run."[96] Planck replied in effect that a better concession to the spirit of the times would have been to say nothing about the author of relativity. Even in 1942 Planck found it prudent not to mention Einstein's name in his contribution to the Nazi film archive, although he praised relativity on camera as "the completion and crown of the whole edifice of theoretical physics"; and the same year he advised Laue not to write "Einstein" in his book on relativity theory. As late as 1943 some Nazi bureaucracies had not yet noticed, or accepted, the official rehabilitation of the work—without the name—of Einstein.[97]

Under these circumstances the content of a talk Planck gave at the Nazi Foreign Officers Club (even there he took his message!) in 1943 or 1944 will appear doubly extraordinary. A Swedish journalist, Gunnar Pihl, was there:[98]

96. Anon. to Neuberg, 23 [Dec.] 1940, Neuberg to Planck, 30 Dec. 1940, and Planck to Neuberg, 10 Nov. 1939 (ANP).

97. Planck to Neuberg, 5 Feb. 1941 (ANP); Planck, "Selbstdarstellung" (1942), 9; Planck to Laue, 3 July 1942 and 5 June 1943 (DM); Direktor, Institut für theoretische Physik, Munich, to Dekan, Naturwissenschaftliche Fakultät, Leipzig, 29 Oct. 1942, in Wendel, in Harig and Mette, eds., *Naturwissenschaft-Tradition-Fortschritt* (1963), 332.

98. Pihl, *Germany* (1944), 228–29.

[Planck] talked about his views of existence. Quietly, humbly, wisely. . . . He mentioned the Jew Einstein as a leader and way-shower in the world of thought, he looked beyond raw prejudices and fanatics, entirely regardless [of where he was]. With his gentle voice . . . he called forth a vision of the divineness of life and its government by law. . . . The little man in black . . . had been too great to be affected by any Nazi efforts at change. . . . It was like being present at a ceremony or a sermon. A violent contrast with the spirit of the place.

The Head Office for Science counseled against Planck's further lecturing.

Planck's lessons on religion, which made no reference to Christ, did not please that religious body the Gestapo, which insisted on hard-selling the Savior in its model sermon for delivery in occupied territory. "Neutrality toward Christ is a dangerous, even an impossible thing." The omission of Christ also interested the Nazis for a subtler reason. It evidently confirmed the rumor, which had been circulating for some years, that Planck's mother was not of Aryan descent. Being scientific, the Hauptamt Wissenschaft quantified the blemish and made Planck one-sixteenth Jewish. Planck tried to quash the rumor by having Hartmann insert in the second edition of his book that Planck's mother came from a family of pastors in East Prussia. In sum, Planck had enemies in the higher reaches of the Nazi bureaucracy.[99] The greatest liar of them all, the Reichsminister für Volksaufklärung und Propaganda (Popular Enlightenment and Propaganda), P. J. Goebbels, considered Planck at the very best to be cool toward the Reich. In a move characteristic of the in-

99. Boberach, ed., *Berichte* (1971), 723–24; Baumgärtner, *Weltanschauungskampf* (1977), 92; Planck to Hartmann, [May 1942] (MPG); HH, 39.

coherence of the regime, he prohibited the city of Frankfurt am Main from awarding Planck its Goethe prize (a "distinctly cultural prize" could not go to a German who championed Einstein) almost at the moment that Hitler was sending Planck best wishes on his eightieth birthday.[100]

To Planck the army and the captive press were merely the means by which he could impart to others considerations that he had found consoling. An eyewitness recalled the effect of *Sinn und Grenzen* on an audience of medical students in Berlin in January 1945. Planck spoke for forty minutes, scarcely glancing at the manuscript before him. Planck himself was the message. "The gentle voice of the scientist perhaps therefore moved us especially. Never before had we experienced such respectful silence in the auditorium."[101]

As Salvager

Planck's preaching brought him increasingly into danger from air raids. At Koblenz in the fall of 1943, bombs interrupted his lecture; in Frankfurt the destruction was so great that he was not permitted to appear; in Kassel he witnessed a "sight out of Hell," the wasting of a city, and spent the night in a shelter.[102] The previous spring he had moved to Rogatz, on the west bank of the Elbe near Magdeburg, to escape the bombings of Berlin. The roof of his house in Grü-

100. Goebbels, diary, May 1943, quoted in LS, 2, 140; Mayor, Frankfurt/Main, to KWG, 26 Oct. 1943, in O'Flaherty, American Association of University Professors, *Bulletin*, 42 (1956), 441 (quote); AH, 103. Cf. Born, in Royal Society of London, *Obituary notices*, 6 (1948), 179–80, and O'Flaherty, *American scientist*, 47 (1959), 73. Planck received the Goethe prize just after the war, to the great satisfaction of Schweitzer, a previous winner: Schweitzer to Planck, 20 Jan. 1946, in Kangro, *Vorgeschichte* (1970), 235.

101. Bachmann and Trummert, *Münchener medizinische Wochenschrift*, 112 (1970), 159; AH, 113.

102. AH, 105–6; Planck to Laue, 23 Sept. and 19 Oct. 1943 (DM).

newald had been damaged. He planned to have it repaired and to return to Berlin early in May 1943. But he could not find workmen, and he continued his rustication, cut off from the University colloquia and the meetings of the Academy that he had continued to attend in the early war years.[103] He was still vigorous in mind and body (that year he climbed a 3,000-meter mountain), still optimistic, still full of encouragement and reassurance for others. But he was about to sustain a series of disasters that robbed him of almost everything.

The first occurred on the night of 15 February 1944. A tremendous air raid against Berlin obliterated the suburb of Grünewald. Not a thing could be saved from Planck's house. So perished not only his library, diaries, and correspondence, but also the last physical evidences of the high culture of the professors who had settled there, close to nature and the railhead, under the Empire.[104]

Planck took the loss of his home, his possessions, all of the carefully preserved and valuable memorials of a long and productive life in stride. He wrote that life in Rogatz was not bad, that he could work (always his recipe for recovery), that he had finished a new lecture, on pseudoproblems of science, which he looked forward to delivering.[105] Then came

103. On the house: Planck to Dora Martin, 17 Mar. 1943 (MPG), to Hildegard Gravemann, 6 Apr. and 18 June 1943 (MPG), and to Laue, 2 and 3 Dec. 1943, 4 Jan. 1944 (DM). On academic life: Planck to Laue, esp. 13 Jan. 1940, 5 Aug. and 20 Dec. 1941, 11 Mar. 1942; he was still concerned with Academy business in June 1944, according to Marga Planck to Laue, 9 June 1944 (DM).

104. Planck to Max Schirmer, 29 Oct. and 22 Nov. 1943 (MPG); Marga Planck to Annamarie Schrödinger, 9 May 1948 (AHQP/92). Cf. Agnes von Harnack, in "Max Planck zum Gedächtnis," PB, 4 (1948), 166; Planck to Hildegard Gravemann, 27 Feb. 1944 (MPG), and to Schirmer, 20 Feb. 1944 (MPG).

105. Planck to Schirmer, 2 Mar. 1944, and to Hartmann, 24 Nov. 1943, 2 May 1944, 3 Jan. 1945 (MPG).

the news that his granddaughter Emma had tried to commit suicide. Marga went to the girl and brought her to a sanitorium. Planck was then too ill to travel or even to write. Still he could console others, like Kippenberg, who had also suffered large losses.[106] All this was but prelude.

Late in 1944 Erwin Planck and his friend from childhood, Ernst von Harnack, were judged guilty of complicity in the attempt to assassinate Hitler and condemned to death. Erwin probably had no part in the plot, but he knew many of the plotters and certainly sympathized with their cause. A draft for a transitional government, prepared in 1940 for discussion by several of the future unsuccessful assassins, had important contributions from his hand. As a high official for military affairs in the late Weimar regime, he had made contacts through whom he had tried to prompt the army to take action against the Nazis in the summer of 1934. According to his testimony to the police, he met the future plotters in the late 1930s and later participated in discussions about how to end the war by diplomacy. He broke off, he said, on realizing that the objective could not be accomplished.[107]

No doubt Planck knew something about his son's compromising discussions. Not only were they very close, but also they were either members or frequent guests of an old club, the Mittwochsgesellschaft für wissenschaftliche Unterhaltung (Wednesday Club for Scientific Recreation), where the plotters congregated.[108] But he could not have

106. Marga Planck to Laue, 22 Apr. and 11 June 1944, the last requesting that Laue draft a preface to a new edition of *Wege*; Planck to Laue, 29 July and 8 Aug. 1944 (DM); Planck to Kippenberg, 5 May 1944 (KA).

107. Peter Hoffmann, *The German resistance* (1977), 27, 110, 178–79; Peter, *Spiegelbild einer Verschwörung* (1961), 290; Planck to Laue, 8 Aug. 1944 (DM), and to Schirmer, 13 Oct. 1944 (MPG).

108. Peter Hoffmann, *The German resistance* (1977), 117; Bachmann and Trummert, *Münchener medizinische Wochenschrift*, *112* (1970), 161. Klaus Scholder, *Die Mittwochs-Gesellschaft* (1982), 368, lists Heisenberg

guessed the price the regime would exact for Erwin's long-standing opposition to it. Planck set "Heaven and Hell in motion," he wrote Laue, to have the sentence commuted, and, as he told Hahn, he had reason to believe that he would succeed.[109] On 18 January 1945 Planck learned that a pardon would soon come; five days later, without warning, against all hope and expectation, without a message or a token from him, Erwin was executed. "No one understands it," Marga wrote. "'They say' that Himmler, who intervened for him, was then at the eastern front and that Hitler was in Berlin."[110] Faceless and empty little men, the sort of people Planck could not reach, killed his son.

The news almost killed Planck. "My sorrow cannot be expressed in words," he wrote Sommerfeld. And, to a niece and nephew: "He was a precious part of my being. He was my sunshine, my pride, my hope. No words can describe what I have lost with him."[111] Among the letters of condolence was one from Kippenberg, who as usual pointed out that there was no prospect of improvement in this life. This time Planck agreed. "You are right. If there is consolation anywhere it is in the Eternal, and I consider it a grace of Heaven that belief in the Eternal has been rooted deeply in me since childhood. . . . God protect and strengthen you

but not the Plancks as a member. The arrest of many of its leaders immediately after the attempt destroyed the club (ibid., 11, 42–43).

109. Planck to Laue, 2 Nov. 1944, in AH, 111, and to Hahn, 5 Nov. 1944 (DM); to the Hans Hartmanns, 3 Jan. 1945 (influential people have been mobilized) (MPG); to Laue, 29 Jan. 1945 (DM); Marga Planck to Laue, 8 Mar. 1945 (DM), and to Frau Otto Hahn, 11 Feb. 1945 (MPG). Cf. Glum, *Wissenschaft, Wirtschaft und Politik* (1964), 395–96, 413, 429.

110. Marga Planck to Laue, 18 Mar. 1945 (DM).

111. Axel von Harnack, in "Max Planck zum Gedächtnis," *PB*, *4* (1948), 170–71; Planck to Sommerfeld, 4 Feb. 1945 (AHQP), and to Fritz and Grete Lenz, 2 Feb. 1945 (MPG), in AH, 113–14, and, in similar words, to Schirmer, 29 Jan. 1945 (MPG).

for everything that still may come before this insanity in which we are forced to live reaches its end." After Erwin's death Planck took no pleasure in life.[112] The insanity grew worse. To distress of spirit was added agony of body. Planck's vertebrae fused, literally doubling him over in pain. Rogätz became a battlefield. Scarcely able to walk, the old man had to leave his sanctuary. He and his wife hid in the woods and slept in haystacks. "The very worst [as Marga Planck recounted it to Hartmann] was the frightful sufferings that Uncle Max had to bear. He often screamed from the pain." At last came rescue by American officers alerted to the Plancks' probable plight by Robert Pohl, professor of experimental physics at Göttingen. Utterly destitute, the Plancks found refuge with a niece in Göttingen, which had not been damaged by the war.[113]

After five weeks in a hospital Planck could again shuffle about, but he was very tired and had trouble concentrating and remembering. The last war years had catapulted him into old age. Still he undertook to lecture on religion and science; his reading still affected his listeners and gave him the satisfaction, in Marga's words, "of still doing something for the general public." His last lecture trip required travel in an unheated railroad car in January. Someone asked him why he did it. "At eighty-nine I cannot be productive scientifically; what remains to me is the possibility of following the advances my work prepared and of responding to the wishes of people struggling for truth and knowledge, especially young people, by repeating my lectures here and there."[114]

112. Planck to Kippenberg, 14 Mar. 1945 (KA); Marga Planck to Annamarie Schrödinger, 10 Nov. 1947 (AHQP/92).
113. Marga Planck to Hartmann, 12 Dec. 1945 (MPG); AH, 114–16; Laue, Nwn, 35 (1948), 7.
114. HH, 27–28, re a lecture in Hamburg; AH, 116–17, 123–24, 126; Marga Planck to Hartmann, 15 May 1947 (MPG, quote), and to Som-

other duty the invitation of the Royal Society of London to participate, as the only German, in a belated commemoration of the three hundredth birthday of Isaac Newton. Planck's name did not appear in the printed program. His appearance caused surprise and a splendid, pathetic incident. We have a report from the scene:

> At the opening session the foreign guests were introduced to the President of the Royal Society. In a loud voice a page, dressed in elaborate uniform, read off the name and nationality of each foreign guest; the guest then shook hands with the President and retired to his seat. At the very end, after all guests listed had been introduced, the page announced, "Professor Max Planck from no country." Planck received an ovation of clapping. Later the President corrected the wording. "The announcement should have been, Professor Max Planck from the world of science."[114a]

The trip tired the body but strengthened the spirit. As he wrote Neuberg, the meeting gave him the good feeling of having contributed a mite to progress. "I am not one of those who let themselves be bitter." When asked, he located the spring of his spirit in the fixed belief, implanted in him in childhood, in "another world, exalted above ours, where we can and will take refuge at any time."[115]

Planck may not have found his religious beliefs confirmed in the next world, but they abundantly proved themselves in this one. At least for him. Some people, disbelieving that

merfeld, 3 July 1946 (AHQP/33); Planck, in *Atlantis, 14* (1947), 14, as quoted by D. Hoffmann, in Urania, *Zur Geschichte* (1982), 25.

114a. Walter R. Miles to Ernest Merritt, 19 Feb 1947, Cornell Univ. Archives, 14/22/46 (P1).

115. Planck to Neuberg, 17 Sept. 1946 (ANP); Planck to A. Berthollet, 28 Mar. and 14 Dec. 1945, in "Max Planck zum Gedächtnis," *PB, 4* (1948), 162; cf. Born, in Royal Society of London, *Obituary notices, 6* (1948), 180.

science can or should be compatible with religion, have criticized his reconciliation as harmful to the one or to the other. On the scientistic side, Marxist writers have lamented such flagrant bourgeois idealism in a man whom otherwise they admire as a self-taught materialist.[116] On the other side, Church spokesmen could scarcely become enthusiastic about Planck's deism, which omitted all reference to established religions and had no more doctrinal content than Einstein's Judaism. It seemed useful therefore to paint the lily, to improve the lesson of Planck's life for the use of proselytizers and to associate the deanthropomorphizer of science with a belief in a traditional Godhead.

Six months before his death from a stroke on 4 October 1947, a rumor started that Planck had converted to Catholicism. An engineer applied to him for the reason that had brought him to such a step. The reply was not useful to missionaries. He had always been deeply religious, Planck said, but he did not believe in "a personal God, let alone a Christian God."[117] A God without qualities, a religion without trappings, life without compartments, knowledge without divisions—in brief, a worldview without extremes—have little appeal to prophets and promoters.

It remained only to bring home the ark. Bosch's successor as president of the Kaiser-Wilhelm-Gesellschaft had taken his own life; many of its institutes had been damaged or destroyed, and their staffs scattered; the Society's assets had fallen to nothing. In the extremity the general secretary, who had moved the seat of administration to Göttingen, asked

116. For example, Jánossy, in Kockel et al., eds., *Max-Planck-Festschrift* (1959), 397; Klohr, in Halle, Univ., *Wissenschaftliche Zeitschrift*, 6:2 (1957), 293–99. Cf. Dinkler, *Zeitschrift für Theologie und Kirche, 56* (1959), 201–2, 216–17. Even the astute Vogel takes Planck's religiosity to have been a "Trojan horse for religion and the Church" (*PhW,* 234).

117. Planck to W. H. Kick, 18 June 1947, in Herneck, *Wissenschaftsgeschichte* (1984), 89–90; *PhW,* 227, lists philosophers and theologians who tried to christianize Planck's religion.

Planck to resume the presidency until the Society could be reconstituted. Under Planck's authority, the surviving directors in the Western zones of occupation chose Otto Hahn as his second successor. Hahn accepted effective 1 April 1946.[118] That by no means put affairs in order. Although the British and the French authorities allowed the Society to continue in their zones, the American occupation forces wished to dissolve it, perhaps from the conviction that it had succumbed to Nazi influence.

The situation called for quiet negotiations based on personal connections, according to Marga Planck, exactly the realm in which her husband had excelled. Planck's advocacy and pressure from the Royal Society of London strengthened the British against the Americans; and the British authorized the incorporation within their zone, which included Göttingen, of a possible successor to the Kaiser-Wilhelm-Gesellschaft, provided its name be changed to something less suggestive of militarism. A suitable name was not far to seek. On 11 September 1946 there came into existence the "Max-Planck-Gesellschaft für die Förderung der Wissenschaft in der Britischen Zone" (MPG for the Advancement of Science in the British Zone). In the argument to persuade the Americans that the Society had not aligned itself with the Nazis, Planck's name and the sacrifice of his son figured prominently. Soon the American authorities made common cause with the British, and in July 1949 the three Western powers approved the statutes of a Max-Planck-Gesellschaft in their zones of occupation.[119] Planck had the satisfaction of

118. AH, 119–20, and Hermann, *New physics* (1979), 111–15; Elisabeth Heisenberg, *Das politische Leben* (1980), 150–54; Planck to Direktoren, 24 July 1945 and 23 Mar. 1946, and to Hahn, 25 July 1945, in MPG, *50 Jahre* (1961), 189–90, 201.

119. Marga Planck to Hartmann, 27 Jan. 1946 (MPG); Gründungssitzung, "Protokoll," 11 Sept. 1946, and Research Control Authorities to Hahn, 8 July 1949, in MPG, *50 Jahre* (1961), 202–5, 231; Stamm, *Zwischen Staat und Selbstverwaltung* (1981), 81–98.

seeing in leading positions in the recreated Society men he had urged to weather the storm in Germany. The high administration included Laue as secretary as well as Hahn as president. The director of the Max-Planck Institute for Physics was Heisenberg, who had taken over from Debye in 1942 against the by then ineffectual protests of the Aryan physicists.[120]

By Their Fruits Ye Shall Know Them

There are those who think that Planck could and should have done more. Had men like him organized a grand protest or, better, a general strike of professors, immediately on learning of the law for the cleansing of the civil service, perhaps the newly established regime would have had to moderate its policies toward Jewish academics. On this theory, conspicuous protest, even if futile or counterproductive, was a moral imperative. Planck had helped to draw up a declaration on this basis against a purge of scientists by the Soviet government in 1930. He added to the draft a reference to the "obligation of conscience" to protest publicly against so evident a violation of "the most fundamental laws of human social life." Planck's "obligation of conscience" was for Einstein a categorical imperative, "an absolute duty." Obeying it might not help the victims of the purge, but it would satisfy the consciences of the signers and alert the rest of the world to the Soviet menace. Would such a course help Western Europe? "That is by no means certain," Einstein wrote. "Even in our case external conditions are developing slowly but steadily in a threatening direction."[121]

When, in 1933, the threat materialized, Einstein made his protest and tried by parable to persuade Planck of the inescapable obligation to speak out against the destruction of the

120. ADB, 188–95; Richter, in Mehrtens and Richter, *Naturwissenschaft, Technik und NS-Ideologie* (1980), 116–41.
121. Einstein to Planck, 10 Oct. 1930 (AE).

Jewish professoriat. Suppose that you were a professor in Prague, he put it to his old friend, when a new regime came to power and deprived Czech teachers of their means of livelihood and refused to allow them to leave the country. "Would you think it right to accept the dismissals silently, without taking their part?"[122] Einstein never forgave Planck for his public silence and even Laue criticized him for not being more "obdurate."[123] But whereas Einstein was a free spirit, who had little attachment to Germany and felt no responsibility for the institutions that supported him there, and Laue was not so highly placed as to suppose that he could influence the course of events, Planck had both responsibility and position, and a rule for exercising them. He decided that it would be more fruitful to stay and salvage rather than to run and gesture. And he had his fruit in the nuanced independence of the Academy and the Kaiser-Wilhelm-Gesellschaft during his tenure of office.

Again, had men like Planck realized earlier that one part of the Nazi bureaucracy could be played off against another they might have won important concessions. Infighting of the kind that toppled Stark indicated what might have been done. Another indication was the belated realization on the part of high Nazi officials that they had made a grave mistake by allowing Rust to ruin the universities. Industrial and military users of high technology eventually joined with rank-and-file scientists and Planck and other white Jews to discredit and neutralize the Aryan physicists.[124] Could not an alliance of scientists and industrialists have exploited cracks in the bureaucracy much earlier?

122. Einstein to Planck, 6 Apr. 1933 (AE).

123. Cf. Gregg, diary, 16 June 1934 (RF 717/2), on Laue; Marga Planck to Einstein, 26 Nov. 1947 (AE), on Einstein's refusal to write Planck after their exchanges in 1933.

124. ADB, 207–8; Ramsauer, *Physik-Technik-Pädagogik* (1949), 82–84; LS, *3*, 132–33, 140; Grundemann, in Dresden, Technische Hochschule, *Wissenschaftliche Zeitschrift*, *16:5* (1967), 1626.

Planck was temperamentally unfit for public protests against constituted authority and for a cynical manipulation of political differences or personal ambitions. He tried to build enclaves within the system, in which relatively normal life and work could be pursued. Here the most evident fruit of his labor was the Kaiser-Wilhelm Institute for Physics. And he tried, through his lecture tours, to draw the attention of others to the enduring values of life and to bring them hope from his almost inexhaustible supply. Here, too, he had some success.

Judgment by fruits, however, is an ambiguous measure. To point up the ambiguity, we need consider only the physics institute for which Planck struggled so long and so hard. He could rightly have taken pride in this ark when launching it in 1936 and when it docked in 1945. But if required to judge it under sail in the early 1940s, he would have wished it at the bottom of the sea. Under Heisenberg, whom he had urged to stay on in Germany, the institute became the center of a project to exploit nuclear energy. This is not the place to discuss the intentions of Heisenberg, but merely to indicate how the apparatus and workplace provided by Planck were used. In 1943 Heisenberg wrote to the administration of the Kaiser-Wilhelm-Gesellschaft that the high-tension set at the institute had been turned into a strong neutron source, "in order to put it to use in the investigation of problems in atomic physics posed by Army Ordnance [Heereswaffenamt]." Many other researches into similar problems had been undertaken by the institute.[125]

Planck had been worried from the beginning about the misuse of nuclear fission. Manfred von Ardenne, who did some work on the uranium project, had a conversation with Planck during a drive in November 1940. A copy of the *Völkischer Beobachter* announcing the "enlightenment" the Luft-

125. Heisenberg to KWG, Generalverwaltung, 22 June 1943 (MPG A/39); Irving, *The German atomic bomb* (1967).

waffe was about to bring to England lay on the back seat of the car. The two agreed that, with the available means, Germany could not win the war. Then Planck said, "I am very worried." Ardenne guessed that he had Hahn's discovery of fission in mind. "It must be used for the good of humanity," Planck remarked. And then: "It will happen otherwise."[126]

It happened to Hiroshima, not to London. Whatever the objectives of Heisenberg's group, the structural weaknesses of the Nazi bureaucracy, its contempt for science, and the lack of disposable industrial resources ruled out making nuclear bombs in wartime Germany. It was not luck, therefore, in Planck's mind, that his ark did not design the torpedos that would sink mankind. Or that the Russians did not get the atomic bomb first, which, in Planck's judgment, would have meant the end of "our culture."[127] It was in the order of things.

Despite all his sad experience, Planck stuck to the fundamental belief that seemed so well founded in the heady days of Germany's rise to greatness. Then intellectual and social life, duty to and respect for the state, and moral imperatives were all of a piece. Subsequently state, science, and morality had gone their separate ways; the world order disappeared from view; and Planck's virtue, knowledge, and authority availed nothing to deflect disaster from those closest to him. But that, too, may have been part of the world order. How to tell? "An event, which we earlier lamented as a misfortune, [may prove to have] protected us from a still greater misfortune; then perhaps our lament will turn into satisfaction and joy."[128]

126. Ardenne, *Mein Leben* (1984), 148–50. British Intelligence rated both Ardenne and Heisenberg as Nazis for their wartime activities (Chadwick Papers 4/56, Churchill College, Cambridge).

127. Planck to Klippenberg, 7 Feb. 1946 (KA).

128. *Wege*, 289 (1936). Cf. Einstein to Marga Planck, 10 Nov. 1947 (AE), and Planck to Neuberg, 16 May 1942 (ANP).

And there is the dilemma of dilemmas. Few actions are mandated by categorical imperative. Moral behavior is guided in practice by maxims derived from social experience and judged by its near and remote consequences. How and when should the reckoning be made? Did Planck take the best way with the Nazis by staying in office and preserving, as best he could, his clear conscience and good will? Did his worldview ennoble or betray him?

AFTERWORD, 2000

Several key documents concerning Max Planck's behavior during the Third Reich have come to hand since *Dilemmas* was first published in 1986. They include letters about his account of his visit to Hitler, other correspondence, and entries in the diaries of Thomas Mann and Lotte Warburg. Warburg's declarations to her diary, published in 1989, provide a convenient framework for an assessment of the new evidence. As the half-Jewish daughter of the full Jew Emil Warburg, Planck's long-time friend and colleague at the University of Berlin, she had known Planck since her childhood. As the sister of Otto Warburg, the cynical Nobel Prize–winner who directed the KWI for Biochemistry, she remained close to gossip about science after her father's death in 1931. She thought she knew how to evaluate Mandarin professors. Here is what she wrote:

> [1933] Planck has become a stone-aged man, stooped, pitiful. I saw him shuffling through the park, untidy and unkempt. I gave him all the news about [firings at] the universities. He knew nothing about it. He said that no one asked him about things anymore, that science was no longer worth anything.
> [1934] Why does he not resign? Why does he stay on and allow himself to discharge people from the KWG? . . .

Why does he go about hunched over, moaning and complaining, instead of throwing back his head and damning them all?
[1935] A senile old man who is losing the remnants of the courage and character he apparently never had.

This last evaluation, written in the safety of Switzerland, was prompted by a speech of Planck's about the Physikalische Gesellschaft in which, as a routine precaution, he made no mention of the important role of Emil Warburg. An equally routine thank-you to Hitler for supporting German science provoked Warburg to even greater disdain:

[1937] Such a monstrous lie! . . . In spite of it all, everyone who utters the name of Planck will say . . . "Ah, Planck, there's an honorable character!" Until the end of his life he will wear the mask of an honorable, selfless, true scientist, faithful to his convictions, and no one will get at the truth, at the monumental cowardice and spinelessness with which his last years were filled. No one![1]

Why did he retain his offices? How could he have made those necessary small obnoxious concessions, the Hitler salute, deference to Nazi leaders, obsequious telegrams to the Führer, swastikas in lecture halls, omissions of the names of Einstein, Warburg, and other Jews from his speeches?[2] What was so precious to him that to have it he could allow himself to preside over the dismissal of Jewish and other "undesirable" colleagues from the KWG and the Academy of Sciences? How did he maintain his reputation as a man of probity, kindliness, and honor in life and also, as Warburg foresaw, after death?

1. Entries of 31 Oct. 1933, 5 Oct. 1934, 28 Jan. 1935, and 17 Oct. 1937, in Warburg, *Eine Närrin* (1989), 187, 235, 250–51, 295–96.
2. Cf. Olff-Nathan, *Science* (1993), 20.

We know the answer to the first set of questions. Planck remained in office largely from a sense of duty owed not to individuals, certainly not to the state, but to the institutions of German science he served. He had come to identify ever more strongly with these institutions, many of which he had nourished from their beginnings. He had been instrumental in defeating the dire prediction of Svante Arrhenius, made in 1922: "The scientific spirit is still very high in Germany, but it is condemned soon to collapse."[3] The return of German science to its rightful place in the world gave Planck two misleading indications: come what may, German science should be preserved and, through the efforts of energetic men of good will like himself, could be preserved. He and they overrated their power and influence.

Otto Warburg made this last point to his sister shortly after receiving notification of his Nobel Prize in 1931. He referred to their father and to Planck among the old professors who exaggerated their importance. Her reply may help in interpreting her later diatribes against Planck. "No," she said, "these old men do not suffer so much from exaggerated importance as from vexation that they can no longer be active in the fields that had given them their orientation in life." She gave their father as an example. He never should have married. He cared only for his work; he insisted on logic and order in everything; he had no humor, no feelings, no sense for the all-too-human. That, she said, was typically Jewish: "Generally speaking, there is no cheerful, radiant Jew, among the great only Einstein, who, however, is no longer a Jew. . . . Einstein, [Max] Liebermann, they are un-Jewish, free, without willpower, happy, giving happiness."[4]

3. Arrhenius to Osterhout, 16 Dec. 1922 (Osterhout Papers, American Philosophical Society, Philadelphia).
4. Warburg, *Eine Närrin* (1989), 115–17 (10 Nov. 1931).

To Lotte Warburg, Planck's character, like her father's and brother's, had too much Jewishness in it.

Both Otto and Lotte Warburg were right. Planck did hold tightly to his science, his image of science, as his technical contributions to it diminished and its demand for influential spokesmen increased. And he exaggerated his influence—not out of vanity, but from an inability to anticipate a regime in which reasoned argument, his only way of exercising influence, would not avail.

There was no way that Planck could have retained the reputation for independence and impartiality that he had brought to the presidency of the KWG once he had decided where duty lay after the Nazi seizure of power.[5] The inevitable loss of self-respect (perhaps the reason for his uncharacteristic uncertainty and untidiness at his chance meeting with Lotte Warburg in October 1933) must have been especially galling when the "resignations" of Jews he had persuaded to stay revealed the futility of his policy. One of them was Carl Neuberg, director of the KWG's Institute for Biochemistry, who for a time invoked the veteran's exemption from the civil service laws. Another was Lise Meitner. Both Neuberg and Meitner bitterly regretted their inertia. Meitner wrote to Otto Hahn after the war: "It is very clear to me today that I committed a great moral wrong by not leaving in 1933, since in the last analysis by staying I supported Hitlerism."[6]

She had put her finger on the fatal flaw of Planck's policy of preservation. As she said, however, her grievance was hindsight. Planck had given his advice in the expectation that the Nazis would have to palliate their policies

5. Macrakis, *Surviving* (1993), 40.
6. Meitner to Otto Hahn, 6 June 1948, in Krafft, *Im Schatten* (1989), 185, also (in English) in Hentschel and Hentschel, *Physics* (1996), 401.

to remain in office. His was by no means the only voice counseling Jewish colleagues to weather the storm for the good of German science. Even exiles, like the physicist Kurt Mendelssohn, who ended up in Oxford, could advise as late as 1936 that anyone not absolutely forced to emigrate should remain in Germany.[7]

In one case Planck's policy of retaining distinguished Jewish scientists as long as possible worked improbably well. Otto Warburg, "the most accomplished biochemist of all time," disliked Germany. He remained there, however, encouraged by Planck, shielded as a veteran, and protected by the Rockefeller Foundation, which had paid for his institute.[8] Warburg managed to keep his place throughout the Second World War apart from an interlude in 1941 when, on the orders of the ministry of education, Planck's successor fired him. Under pressure, two officials of the Reich Chancellery, who with characteristic Nazi schizophrenia simultaneously presided over the murder of 50,000 Jews and other misfits, "aryanized" Warburg, perhaps because the Nazis had been persuaded that he might produce a cure for cancer.[9] He piloted his institute, which did no war work, safely into the harbor of the Max-Planck-Gesellschaft.

Lotte Warburg's prophecy that no one would try to pull away Planck's "mask of an honorable, selfless, true scientist, faithful to his convictions," proved false. Recent historians have found reasons to accuse him of cowardice and collaboration. His euphemistic references to dis-

7. Macrakis, *Surviving* (1993), 65; Sime, *Meitner* (1996), 204–207; Mendelssohn to Hugo Gutsche, 6 Feb. 1936 (Mendelssohn Papers, Bodleian Library, Oxford); supra, 153–54.

8. "The greatest biochemist," in *Dictionary of scientific biography*, s.v. "Warburg, Otto."

9. Macrakis, *Surviving* (1993), 64; Warburg, *Eine Närrin* (1989), 300.

missals in his annual reports to the KWG ("Dr. Stern accepted a call from the University of Rochester") have received special, and perhaps undue, attention as illustrations of his "readiness . . . to subordinate [himself] to the regime almost without protest."[10] The most damaging criticism, developed by Herbert Mehrtens, Helmuth Albrecht, and others, asserts that Planck's conservative nationalism ineluctably led him to collaboration with, and manipulation by, the National Socialist regime once he had decided to try to weather the storm in office.[11] They develop their position around the brief notice "My Visit to Hitler," published under Planck's name a few months before his death. The notice reads in full:

> After Hitler came to power I had the task as president of the KWG to wait upon the Führer. I thought to use the opportunity to say a word in favor of my Jewish colleague Fritz Haber, without whose process for making ammonia from atmospheric nitrogen the First World War would have been lost from the beginning. Hitler answered me in these words: "I have nothing against the Jews themselves. But Jews are all communists, and they are my enemies, against them I wage war." To my remark that there are different sorts of Jews, some valuable for mankind and others worthless, and among the first old families with the best German culture, and that distinctions must be made, he replied: "That is not right. A Jew is a Jew; all Jews stick together like leeches. Where there is one Jew, other Jews of all sorts immediately gather. It was up to the Jews themselves to sepa-

10. Deichmann and Müller-Hill, in Renneberg and Walker, *Science* (1994), 163.
11. Albrecht, in Albrecht, *Naturwissenschaft* (1993), 53, 63, followed by Szöllösi-Janze, *Haber* (1998), 658–62.

rate the various kinds. Since they have not done so I must go after all Jews in the same way." To my remark that it would be self-mutilation to make valuable Jews emigrate since we need their scientific work, which, moreover, would primarily benefit foreign countries, he made no response. Instead he uttered some common-places and finished up: "People say that I suffer from a weakness of nerves. That is a slander. I have nerves of steel." So saying, he hit himself hard on the knee, spoke faster and faster, and flew into such a rage that I could only remain silent and withdraw.[12]

This dramatic account, as Albrecht has discovered, was composed by Marga Planck owing to her husband's increasing physical frailty. The Plancks evidently thought it a duty to comply with the request of the editor of the *Physikalishe Blätter* for a contribution to its program of presenting the actions of German physicists during the Third Reich in the best possible light. Albrecht notes that Marga's draft divided Jews into "Western" and "East-ern," not "valuable" and "worthless," a categorization that, she observed, "might perhaps, rightly, be attacked"; "not all Eastern Jews are worthless." Albrecht notes fur-ther that Planck almost certainly would have based his appeal to Hitler on a distinction in geography rather than on merit, since Hitler had already scotched division by degrees of goodness. The distinction between Western and Eastern, or assimilated and unassimilated Jews, was standard in Germany, indeed, in much of the world, between the wars. Preferring the one to the other was a form of anti-Semitism, since every Jew bore the burden of proof that he (to quote from Ernest Lawrence of the University of California, which prided itself on its rela-

12. Planck, *PB, 1* (May 1947), 143; also in Hentschel and Hentschel, *Physics* (1996), 359–61.

tive absence of anti-Semitism) "has none of the characteristics that some non-Aryans have."[13] The distinction between worthless and valuable, on the other hand, accurately reflects Planck's private utilitarian calculus: efforts at salvage should be directed, firstly and obviously, at people good at science.

Having undermined the reliability of "Planck's" account of his tête-à-tête with the Führer, Albrecht accepts the authority of a letter written two weeks after the interview by Werner Heisenberg. According to Heisenberg, Hitler had assured Planck that "the government will do nothing beyond the new civil service law that could hurt our science."[14] Heisenberg did not mention Jews or rages. Albrecht infers that Planck's main purpose in seeing Hitler was to assure him of the KWG's fidelity in return for the privilege of achieving *Gleichschaltung* on its own. Alignment was to require the dismissal of some eighty employees. State contributions to the KWG's budget increased in consequence.[15] Will he nil he, Albrecht concludes, Planck made himself a collaborator.[16]

Whatever Planck's degree of collaboration, Heisenberg's letter cannot be accepted as a reliable guide to what transpired during the conversation with Hitler. Heisenberg did not describe the interview but merely referred to it to help persuade his correspondent, his former teacher Max Born, not to emigrate. His letter to Born, who was not a member of the KWG, does not support the claim

13. Albrecht, in Albrecht, *Naturwissenschaft* (1993), 457–59; Heilbron and Seidel, *Lawrence* (1989), 248–51 (quote).

14. Albrecht, in Albrecht, *Naturwissenschaft* (1993), 47; Heisenberg to Born, 2 June 1933, supra, 154. The date of the interview was 16 May 1933.

15. Albrecht and Hermann, in Vierhaus and vom Brocke, *Forschung* (1990), 359–82; Macrakis, *Surviving* (1993), 65–67, 71–72.

16. Albrecht, in Albrecht, *Naturwissenschaft* (1993), 52–53; Mehrtens, in Mehrtens and Richter, *Naturwissenschaft* (1980), 32–33.

that Planck sold himself to Hitler for the promise of a special immunity for the Society. Other contemporary reports of the interview support the view that Planck dared to raise the question of ameliorating the plight of "valuable," Westernized Jews. In her diary for 17 July 1933, Lotte Warburg recorded a visit from the wife of Erwin Schrödinger. According to Mrs. Schrödinger, "Hitler . . . said to Planck that he was not an anti-Semite, as people always label him; he is only against communism, but the Jews have all become communists. That is the only reason to fight them. Planck had the impression that Hitler is now very tired of the entire Jewish business, but that he cannot stop it."[17]

Thomas Mann recorded in his diary for 2 June 1933 that Hitler had told Planck that his only quarrel with the Jews was their "nasty identification with Marxism." Six months later Mann's friend the Jewish novelist Jacob Wassermann visited him in exile in Switzerland. Wassermann was able to describe the interview in detail:

[Planck] returned home completely crushed. It was like an old brewer's wife talking mathematics, a quarter-education with obsessive ideas, more hopeless than anything the famous scientist and thinker had ever heard in his life. Two worlds come together as a result of the one's rise to power: the world of knowledge, erudition, and disciplined thought must attend to the arrogant, dogmatic expectorations of a revolting dilettantism, bow, and withdraw.[18]

These extracts do not suggest that Planck left Hitler pleased with himself at having obtained relief for the

17. Warburg, *Eine Närrin* (1989), 170.

18. Mann, *Tagebücher, 1933–1934* (1977), 101, 272, which differs somewhat from the translation in Mann, *Diaries* (1983), 161 (2 June 1933), 186 (20 Dec. 1933); Koestler, *Wassermann* (1996), 80–83.

KWG. More information about his mood comes from a letter he wrote to his former student Gabriele Rabel a week after the interview: "In these times you count yourself lucky to retain any optimism or buoyancy."[19] Rabel came from a wealthy, cultured family of "valuable" Jews. One of her relatives, Ernst Rabel, a world-famous jurist, directed the KWG's Institute for Civil Law. Planck helped Rabel to find her calling ("literature and philosophy suit your versatility and originality better than the soberer subject of physics, which . . . sets tight fetters on the creative imagination)," gave her a letter of introduction to colleagues abroad, where she spent the war, and—it was one of those necessary compromises—eventually had to fire Ernst.[20]

As Lotte Warburg foretold, Planck retained his reputation for probity, in death as in life. In the summer of 1947, Marga described to an American professor the way she thought that Planck should and would be remembered:

> How honored and loved my husband is we have again experienced during his illness. And . . . that goes not only for the scientist, but even more for his human qualities: the absolute purity and probity, the fairness and limitless goodness that people close to him see every day.[21]

And not only people close to him. Gerhart Hauptmann, winner of the Nobel Prize in literature, a close contem-

19. Planck to Rabel, 24 May 1933 (Burndy Library, Smithsonian Institution).

20. Planck to Rabel, 22 July (quote) and 1 Nov. 1927 (Burndy Library, Smithsonian Institution); Rabel, *Kant* (1963), 54n; Strauss and Röder, *Dictionary* (1983), *2*, 934.

21. Marga Planck to Ernest Merritt, 6 Mar. and 13 June (quote) 1947; W. R. Miles to Merritt, 19 Feb. 1947 (MS 14/22/46, Cornell University Archives).

porary of Planck's, had kept his distance. He knew Planck through his general essays on science. In 1942 he learned from reading one of them about the possibility of creating a nuclear explosion capable of annihilating the world. This frightful prospect duly appeared in a novel Hauptmann did not live to finish.[22] Early in 1946, close to death, he wrote to Planck:

> For decades I have silently held you in reverence. Your writings have shown me, as no others did, how one comes to science. My love has been mute. . . . So now, at the end of my life, I will no longer put off expressing my unlimited love and admiration to my chancellor and teacher Planck.[23]

Hauptmann may be taken to represent the National Socialist view. He was a fellow traveler of the regime, a Nazi sympathizer, an intellectual celebrated by Goebbel's propaganda machine. According to his biographer, "He gave in to temptation, remained passive, and symbolized the defenselessness of the German educated middle classes against fascism."[24] Did Hauptmann see the same qualities in Planck? If so, he did not share the view of the Allied forces of occupation, which made Planck the exorcist of the bad spirits of the evil-sounding name Kaiser Wilhelm; of the communist regime in the Deutsche Demokratische Republik, which elevated Planck to double hero status, once for joining Lenin's crusade against Ernest Mach and again for resisting, however modestly, the anti-communist Nazis; and of former Jewish associates, like Meitner, who kept alive the image of the man of

22. Hilscher, *Hauptmann* (1988), 453–54. The unfinished novel is "Der neue Christophorus."

23. Hauptmann to Planck, 31 Jan. 1946 (Hauptmann Papers, Preussesches Kultur-Besitz).

24. Hilscher, *Hauptmann* (1988), 403–22 (quote).

exceptional probity, high ideals, and wide culture they had known before 1933.

The character is now established. A historian speaking at the MPG's celebration of the fiftieth anniversary of its eponym's death explained why Planck will not be stripped of his mask, whether or not he wore one:

> Like politics, science requires its impressive symbolic figures to bring its message effectively to a general public. Max Planck was and remains such a figure. He stands for professional excellence and the sustained search for truth, for scientific and personal integrity, for humanity and truthfulness, humility and modesty, and not least for deep engagement with science policy and organization of the scientific enterprise.[25]

The KWG's anniversary celebrations featured a filmed interview of the symbolic figure made by the Goebbel's machine in 1942. As propaganda it does not rank high. Planck gave a few facts about his background and career, about the teachers who inspired him, and about his modest ambition to study without rewriting the laws of physics. He indicated his reluctant acquiescence in his own great discovery, the quantum, "because it contradicts all the ideas of classical atomism." And he described his enthusiasm over relativity, "which crowned and completed the entire edifice of theoretical physics"—without mentioning his old friend Einstein. As for quantum theory in the form given it by Heisenberg and Schrödinger, Planck agreed with the unmentionable Einstein that it was far from a finished theory. That did not bother him much:

> To some extent it is unsatisfactory [he conceded] but on the other hand it is proper and gratifying, for we

25. Hoffmann, in MPG, *Planck* (1997), 54.

will never come to the end, to finality. Scientific work will never stop, and it would be terrible if it did. If there were no more problems, you would put your hands in your pockets and your head on a pillow and would work no more. In science rest is stagnation, rest is death.[26]

For Planck work was a duty, an opiate, and a salvation.

In his filmed interview, Planck spoke extemporaneously, without the careful preparation that advancing age made necessary to ensure that he said what he wanted to convey: "Whoever wants to know who I am I ask to read what I have written and printed."[27] It is not enough. We must also know what he did in obedience to the things he wrote. What he did during the Nazi period was to act in accordance with a worldview that allowed him no escape from his situation with his honor intact. That was the saddest of his catastrophes. The answer to the question with which *Dilemmas* ends is that his worldview betrayed him.

26. Planck, in MPG, *Planck* (1997), 91–94.
27. Ibid., 94.

Works Cited

Albrecht, Helmuth. "Max Planck: 'Mein Besuch bei Adolph Hitler.' Anmerkungen zum Wert einer historischen Quelle." In Albrecht, ed., *Naturwissenschaft und Technik in der Geschichte*. Stuttgart: GNT, 1993. Pp. 41–63.

Albrecht, Helmuth, and Armin Hermann. "Die Kaiser-Wilhelm-Gesellschaft im Dritten Reich." In Vierhaus and vom Brocke, *Forschung* (1990), 356–406.

Deichmann, Ute, and Benno Müller-Hill. "Biological research at universities and Kaiser Wilhelm institutes in Nazi Germany." In Renneberg and Walker, *Science* (1994), 160–83.

Heilbron, J. L., and R. W. Seidel. *Lawrence and his laboratory*. Berkeley: University of California Press, 1989.

Hentschel, Klaus, and Ann M. Hentschel, eds. and trans. *Physics and National Socialism: An anthology of primary sources*. Basel: Birkhäuser, 1996.

Hilscher, Eberhard. *Gerhart Hauptmann*. Frankfurt am Main: Athenäum, 1988.

Hoffmann, Dieter. "Max Planck, 'in weite Kreise hinaus zu popularisieren.'" In MPG, *Planck* (1997), 54–55.

Koestler, Rudolf. *Jakob Wassermann*. Berlin: Morgenbuch, 1996.

Krafft, Fritz. *Im Schatten der Sensation. Leben und Wirken Fritz Strassmann*. Weinheim: Verlag Chemie, 1981.

Macrakis, Christie. *Surviving the swastika: Scientific research in Nazi Germany*. Oxford: Oxford University Press, 1993.

Mann, Thomas. *Tagebücher, 1933–1934*. Ed. Peter de Mendelssohn. Frankfurt am Main: S. Fischer, 1977.

———. *Diaries, 1918–1939*. Trans. Richard and Clara Winston. London: A. Deutsch, 1983.

Mehrtens, Herbert. "Das 'Dritte Reich' in der Naturwissenschaft." In Mehrtens and Steffen Richter, eds., *Naturwissenschaft, Technik und NS-Ideologie*. Frankfurt am Main: Suhrkamp, 1980. Pp. 15–87.

MPG (Max-Planck-Gesellschaft). *Max Planck. Vorträge und Ausstellung zum 50. Todestag.* Munich: MPG, 1997.

Olff-Nathan, Josiane. "Introduction." In Olff-Nathan, ed., *La science sous le Troisième Reich.* Paris: Seuil, 1993. Pp. 7–29.

Planck, Max. "Mein Besuch bei Hitler." *Physikalishe Blätter, 1* (May 1947), 143.

———. "Film: Geheimrat Max Planck (1942)." In MPG, *Planck* (1997), 91–94.

Rabel, Gabriele. *Kant.* Oxford: Clarendon Press, 1963.

Renneberg, Monika, and Mark Walker, eds. *Science, technology and National Socialism.* Cambridge: Cambridge University Press, 1994.

Sime, Ruth Lewin. *Lise Meitner.* Berkeley: University of California Press, 1996.

Strauss, Herbert A., and Werner Röder. *International biographical dictionary of central European emigrés 1933–1945.* 3 vols. Munich: K.G. Saur, 1983.

Szöllösi-Janze, Margit. *Fritz Haber. Eine Biographie.* Munich: Beck, 1998.

Vierhaus, Rudolf, and Bernhard vom Brocke, eds. *Forschung im Spannungsfeld von Politik und Gesellschaft. Geschichte und Struktur der Kaiser-Wilhelm/Max-Planck Gesellschaft.* Stuttgart: Deutsche Verlags-Anstalt, 1990.

Warburg, Lotte. *Eine vollkommene Närrin durch meine Gefühle. Aus den Tagebüchern der Lotte Warburg.* Ed. Wulf Rüskamp. Bayreuth: Verlag Druckhaus, 1989.

BIBLIOGRAPHY

Abb, Gustav, ed. *Aus fünfzig jahren deutscher Wissenschaft*. Berlin: de Gruyter, 1930.

Akademie der Wissenschaften, Berlin. *Max Planck in seinen Akademie-Ansprachen*. Berlin: Akademie-Verlag, 1948.

Akademie der Wissenschaften, Berlin. Institut für Theorie, Geschichte und Organisation der Wissenschaft. *Die Entwicklung Berlins als Wissenschaftszentrum, 1870–1930, 1* (1981), and *5* (1984). (*Kolloquien, 24, 35.*)

Andrade, E. N. da C. "Max Planck." *Nature, 161* (1948), 284.

Apelt, Willibalt. *Geschichte der Weimarer Verfassung*. Munich: Biederstein, 1946.

Ardenne, Manfred von. *Mein Leben für Forschung und Fortschritt*. 7th ed. Munich: Nymphenburger, 1984.

Bachmann, D., and W. Trummert. "Max Planck in der Vorlesung von Sauerbruch." *Münchener medizinische Wochenschrift, 112* (23 Jan. 1970), 158–61.

Badash, Lawrence. "British and American views of the German menace in World War I." Royal Society of London, *Notes and records, 34* (1979), 91–121.

Bader, Joseph. *Forschung und Forschungsinstitute. Eine Monographie der technisch-wissenschaftlichen Forschungseinrichtungen. I. Der Staat als Forscher*. Munich: Verlag der deutschen Technik, 1941.

Balmer, H. "Planck und Einstein beantworten eine wissenschaftliche Rundfrage." *PB, 25* (1969), 558.

Basler, Werner. "Zur politischen Rolle der Berliner Universität im ersten imperialistischen Weltkrieg 1914 bis 1918." *Berlin Universität. Wissenschaftliche Zeitschrift, Gesellschafts- und Sprachwissenschaftliche Reihe, 10* (1961), 181–203.

Baumgärtner, Raimund. *Weltanschauungskampf im Dritten Reich. Die Auseinandersetzung der Kirchen mit Alfred Rosenberg.* Mainz: Matthias-Grünewald-Verlag, 1977.

Bavink, Bernhard. "Die Naturwissenschaften im Dritten Reich." *Unsere Welt, 25* (1933), 225–36.

Benz, Ulrich. *Arnold Sommerfeld. Lehrer und Forscher an der Schwelle zum Atomzeitalter, 1868–1951.* Stuttgart: Wissenschaftliche Verlagsgesellschaft, 1975.

Bergsträsser, Ludwig. *Geschichte der politischen Parteien in Deutschland.* 9th ed. Munich: Olzog, 1955.

Bernhard, C. G., et al., eds. *Science, technology and society in the time of Alfred Nobel.* Oxford: Pergamon, 1982.

Beyerchen, Alan D. *Scientists under Hitler.* New Haven: Yale University Press, 1977.

Bielfeldt, Hans-Holm, et al., eds. *Deutschland, Sowjetunion. Aus fünf Jahrzehnten kultureller Zusammenarbeit.* Berlin: Humboldt-Universität, 1966.

Blackmore, John T. *Ernst Mach. His work, life, and influence.* Berkeley and Los Angeles: University of California Press, 1972.

Boberach, Heinz, ed. *Germany. Reichssicherheitshauptamt. Sicherheitsdienst. Berichte des SD und der Gestapo über Kirchen und Kirchenvolk in Deutschland.* Mainz: Matthias-Grünewald-Verlag, 1971.

Börnstein, R. "Physikalische Unterrichtsübungen für künftige Lehrer." *ZPCU, 19* (1906), 355–56.

Bohr, Niels. "The quantum postulate and the recent development of atomic theory." In Congresso Internazionale dei Fisici, *Atti* (1928), *2,* 565–88.

———. *Collected works.* Ed. L. Rosenfeld et al. Amsterdam: North Holland, 1972–.

Born, Max. "Max Karl Ernst Ludwig Planck, 1858–1947." Royal Society of London. *Obituary notices*, *6* (1948), 161–88.

——. "Die Physik und die Ismen." *PB*, *16* (1960), 147–48.

——. *My life. Recollections of a Nobel laureate*. New York: Scribners, 1978.

Bose, D. M. "Max Planck." *Science and culture*, *13* (1947), 237–42.

Brillouin, Léon. *Vie, matière et observation*. Paris: Albin Michel, 1959.

Brocke, Bernhard vom. "Wissenschaft und Militarismus." W. M. Calder III, Hellmut Flashar, and Theodor Lindken, eds. *Wilamowitz nach 50 Jahren*. Darmstadt: Wissenschaftliche Buchgesellschaft, 1985. Pp. 649–719.

Brüche, Ernst. "Deutsche Physik und die deutschen Physiker." *PB*, *2* (1946), 232–36.

Burchardt, Lothar. *Wissenschaftspolitik im Wilhelminischen Deutschland: Vorgeschichte, Gründung und Aufbau der Kaiser-Wilhelm-Gesellschaft zur Förderung der Wissenschaften*. Göttingen: Vandenhoeck and Ruprecht, 1975.

Clark, Ronald. *Einstein, the life and times*. New York: World, 1971.

Cock, A. G. "Chauvinism in science: the International Research Council, 1919–1926." Royal Society of London. *Notes and records*, *37* (1983), 249–88.

Congresso Internazionale dei Fisici. *Atti*. 2 vols. Bologna: Zanichelli, 1928.

Crawford, Elisabeth. "Arrhenius, the atomic hypothesis, and the 1908 Nobel prizes in physics and chemistry." *Isis*, *75* (1984), 503–22.

——. *The beginnings of the Nobel institution: the science prizes, 1901–1915*. Cambridge: Cambridge University Press, 1984.

Daisomont, M. *Max Planck et la philosophie religieuse*. 2nd ed. Bruges: Editions "de Temple," 1948.

Davidis, Michael, ed. *Wissenschaft und Buchhandel. Der Verlag*

Julius Springer und seine Autoren. Munich: Deutsches Museum, 1985.

Davidson, Martin. *Free will or determinism*. London: Watts, 1937.

Dingler, Hugo. *Die Methode der Physik*. Munich: E. Reinhardt, 1938.

――――. "Determinismus oder Indeterminismus. (Zu Max Plancks gleichnamigem Vortrag." *Zeitschrift für die gesamte Naturwissenschaft*, 5 (1939), 42–57.

――――. "Zur Entstehung der sogen. modernen theoretischen Physik." Ibid., *4* (1939), 329–41.

――――. *Max Planck und die Begründung der sogenannten modernen theoretischen Physik*. Berlin: Ahnenerbe-Stiftung Verlag, 1939.

Dinkler, Erich. "Max Planck und die Religion." *Zeitschrift für Theologie und Kirche*, 56 (1959), 201–23.

Dugas, René. *La théorie physique au sens de Boltzmann*. Neuchâtel, Switzerland: Editions du Griffon, 1959.

Eagle, Albert. *The philosophy of religion versus the philosophy of science*. Manchester: for the author, 1935.

Einstein, Albert. "Max Planck als Forscher." *Nwn, 1* (1913), 1077–79.

――――. "Leo Arons als Physiker." *Sozialistische Monatshefte*, 1919: 25/6, 1055–56.

Einstein, Albert, and Michele Besso. *Correspondance, 1903–1955*. Ed. Pierre Speziali. Paris: Hermann, 1972.

Einstein, Albert, and Max Born. *Briefwechsel, 1916–1955*. Munich: Nymphenburger, 1969.

Einstein, Albert, and Arnold Sommerfeld. *Briefwechsel*. Ed. Armin Hermann. Basel and Stuttgart: Schwabe, 1968.

Erdmann, Benno. *Über den modernen Monismus*. Berlin: Norddeutsche Buchdruckerei, 1914. (Speech at the celebration of the birthday of the kaiser, 27 Jan. 1914.)

Bibliography · 225

Erickson, Ralph W. "Planck's concept of causality." *Journal of philosophy*, *35* (1938), 208–11.

Ferguson, Allan. "Prof. Planck and the principle of causality in physics." *Nature*, *130* (1932), 45–48.

Feuer, Lewis S. *Einstein and the generations of science*. New York: Basic Books, 1974.

Feyerabend, Paul K. "Mach's theory of research and its relation to Einstein's." *Studies in history and philosophy of science*, *15* (1984), 1–22.

Fischer, Karl T. "Vorschläge zur Hochschulausbildung der Lehramtskandidaten für Physik." *ZPCU*, *20* (1907), 65–78.

Forman, Paul. *The environment and practice of atomic physics in Weimar Germany*. Ph.D. thesis, University of California, Berkeley, 1967.

——. "Weimar culture, causality and quantum theory, 1918–27: Adaptation by German physicists and mathematicians to a hostile intellectual environment." *HSPS*, *3* (1971), 1–115.

——. "Scientific internationalism and the Weimar physicists: The ideology and its manipulation in Germany after World War I." *Isis*, *64* (1973), 151–80.

——. "Financial support and political alignment of physicists in Weimar Germany." *Minerva*, *12* (1974), 39–66.

——. "*Kausalität, Anschaulichkeit,* and *Individualität,* or How cultural values prescribed the character and the lessons ascribed to quantum mechanics." In Nico Stehr and Volker Maja, eds., *Society and knowledge*. New Brunswick and London: Transaction Books, 1984. Pp. 333–47.

Forman, Paul, J. L. Heilbron, and Spencer Weart. "Physics *circa* 1900: Personnel, funding, and productivity of the academic establishments." *HSPS*, *5* (1975), 1–185.

Frank, Philipp. "Die Bedeutung der physikalischen Erkenntnistheorie Machs für das Geistesleben der Gegenwart." *Nwn*, *5* (1917), 65–72.

——. "Was bedeuten die gegenwärtigen physikalischen Theorien für die allgemeine Erkenntnislehre?" *Erkenntnis, 1* (1930–31), 126–57.

———. "Zeigt sich in der modernen Physik ein Zug zu einer spiritualistischen Auffassung?" *Erkenntnis, 5* (1935), 65–80.

———. *Einstein. His life and times.* New York: Knopf, 1947.

Fricke, Dieter. "Zur Militarisierung des deutschen Geisteslebens im Wilhelminischen Kaiserreich: Der Fall Leo Arons." *Zeitschrift für Geschichtswissenschaft, 8* (1960), 1069–1107.

Fricke, Dieter, ed. *Die bürgerlichen Parteien in Deutschland. Handbuch der Geschichte der bürgerlichen Parteien.* 2 vols. Leipzig and Berlin (West), 1968–70.

Fuchs, Franz. *Der Aufbau der Physik im Deutschen Museum, 1905–1953.* Munich: Oldenburg, 1957. (Deutsches Museum. *Abhandlungen und Berichte,* 25:3.)

Gehrcke, E. "Wie die Energieverteilung der schwarzen Strahlung in Wirklichkeit gefunden wurde." *Physikalische Zeitschrift, 37* (1936), 439–40.

Gerhards, Karl. "Zur Kontroverse Planck-Mach." *VWPS, 36* (1912), 19–68.

Gerlach, Walter. *Die Quantentheorie. Max Planck, sein Werk und seine Wirkung.* Bonn. Universitäts-Verlag, 1948.

Gesellschaft deutscher Naturforscher und Ärzte. Unterrichtskommission. *Gesamtbericht.* Ed. A. Gutzmer. Leipzig and Berlin: Teubner, 1908.

Glaser, Ludwig. "Juden in der Physik: Jüdische Physik." *Zeitschrift für die gesamte Naturwissenschaft, 5* (1939), 272–75.

Glum, Friedrich. "Zehn Jahre Kaiser-Wilhelm-Gesellschaft zur Förderung der Wissenschaften." *Nwn, 9* (1921), 293–300.

———. "Die Kaiser-Wilhelm-Gesellschaft zur Förderung der Wissenschaften." In Harnack, *Handbuch* (1928), 11–37.

———. *Zwischen Wissenschaft, Wirtschaft und Politik. Erlebtes und Erdachtes in vier Reichen.* Bonn: Bouvier, 1964.

Goldberg, Stanley. "The Abraham theory of the electron: the

symbiosis of theory and experiment." *Archive for history of exact science*, 7 (1971), 7–25.

————. "Max Planck's philosophy of nature and his elaboration of the special theory of relativity." *HSPS*, 7 (1976), 125–60.

Grau, Conrad. "Die deutschen Universitäten und die 200-Jahr-Feier der Akademie der Wissenschaften der UdSSR 1925." In Bielfeldt et al., eds., *Deutschland, Sowjetunion* (1966), 172–78.

Grimsehl, Ernst. "Über den Hochschulunterricht für künftige Lehrer der Physik." *ZPCU, 20* (1907), 1–6.

Groos, Helmut. *Die Konsequenzen und Inkonsequenzen des Determinismus*. Munich: Reinhardt, 1931.

Grumbach, S. *Das annexionistische Deutschland*. Lausanne: Payot, 1917.

Grundemann, S. "Zum Boykott der deutschen Wissenschaft nach dem ersten Weltkrieg." Dresden. Technische Hochschule. *Wissenschaftliche Zeitschrift, 14:3* (1965), 799–806.

————. "Das moralische Antlitz der Anti-Einstein Liga." Ibid., *16:5* (1967), 1623–26.

Haas, Arthur Eric. *Materiewellen und Quantenmechanik*. Leipzig: Akademie Verlagsgesellschaft, 1928.

Haas-Lorentz, G. L. de. *H. A. Lorentz. Impressions of his life and work*. Amsterdam: North Holland, 1957.

Haber, Fritz. "Zum fünfundsiebzigsten Geburtstage des Präsidenten der Kaiser-Wilhelm-Gesellschaft Max Planck." *Nwn*, *21* (1933), 293.

Haber, Fritz, and Otto Hahn. "Richtlinien für die Tätigkeit des Japanausschusses der Notgemeinschaft der deutschen Wissenschaft." *Nwn*, *11* (1923), 31–32, 210–11.

Haberditzl, Werner. "Der Widerstand deutscher Naturwissenschaftler gegen die 'Deutsche Physik' und andere faschistische Zerrbilder der Wissenschaft." In Harig and Mettle, eds., *Naturwissenschaft-Tradition-Fortschritt* (1963), 320–26.

Hahn, Otto. "Einige persönliche Erinnerungen an Max Planck." *MPG. Mitteilungen*, 1957, 243–46.

———. "Zur Erinnerung an die Haber-Gedächtnisfeier vor 25 Jahren, am 29. Januar 1935." MPG. *Mitteilungen*, 1960, 3–13.

———. *My life*. New York: Herder and Herder, 1970.

Hale, Oron J. *The captive press in the Third Reich*. Princeton University Press, 1934.

Harig, Gerhard, and Alexander Mette, eds. *Naturwissenschaft-Tradition-Fortschritt*. Berlin: Deutscher Verlag der Wissenschaften, 1963.

Harnack, Adolf von. *An der Schwelle des dritten Kriegsjahrs*. Berlin: Wiedemannsche Buchhandlung, 1916.

———. "Friedensaufgaben und Friedensarbeit." In Harnack, *Erforschtes* (1923), 279–97.

———. *Die Institute und Unternehmungen der Kaiser-Wilhelm-Gesellschaft zur Förderung der Wissenschaften*. Berlin: Reichsdruckerei, 1917.

———. "Offener Brief an Herrn Clemenceau." In Harnack, *Erforschtes* (1923), 303–5.

———. "Die Kaiser-Wilhelm-Gesellschaft im Jahr 1922." In Harnack, *Erforschtes* (1923), 243–52.

———. *Erforschtes und Erlebtes*. Giessen: Töpelmann, 1923.

Harnack, Adolf von, ed. *Handbuch der Kaiser-Wilhelm-Gesellschaft*. Berlin: Reimar Hobbing, 1928.

Hartmann, Hans. *Max Planck als Mensch und Denker*. 4th ed. Basel, Thun, and Düsseldorf: Ott, 1953.

———. "Max Planck im Kampf um seine Grundideen." *Naturwissenschaftliche Rundschau*, *11* (1958), 128–31.

Hartmann, Max. *Gesammelte Vorträge und Aufsätze*. 2 vols. Stuttgart: G. Fischer, 1956.

Heilbron, J. L. "*Fin-de-siècle* physics." In Bernhard et al., *Science, technology, and society* (1982), 51–73.

———. "The Copenhagen spirit in quantum physics and its earliest missionaries." *Revue d'histoire des sciences* (in press).

Heisenberg, Elisabeth. *Das politische Leben eines Unpolitischen. Erinnerungen an Werner Heisenberg.* Munich: Piper, 1980.

Heisenberg, Werner. "Über den anschaulichen Inhalt der quantentheoretischen Kinematik und Mechanik." *Zeitschrift für Physik, 43* (1927), 172–98.

―――. "Max Planck: Positivismus und reale Aussenwelt." *Zeitschrift für angewandte Chemie, 44* (1931), 189.

―――. "Max Planck: Wege zur physikalischen Erkenntnis." *Nwn, 21* (1933), 608.

―――. *Physics and beyond. Encounters and conversations.* New York: Harper and Row, 1971.

Heller, K. D. *Ernst Mach: Wegbereiter der modernen Physik.* Vienna: Springer, 1964.

Hermann, Armin. "Einstein auf der Salzburger Naturforscherversammlung 1909." *PB, 25* (1969), 433–36.

―――. *Frühgeschichte der Quantentheorie (1899–1913).* Mosbach in Baden: Physik Verlag, 1969.

―――. "Max Planck—praeceptor physicae." *PB, 29* (1973), 483–87.

―――. *Max Planck in Selbstzeugnissen und Bilddokumenten.* Reinbeck: Rowohlt, 1973.

―――. *The new physics. The route into the atomic age. In memory of Albert Einstein, Max von Laue, Otto Hahn, Lise Meitner.* Bonn and Bad Godesberg: Inter Nationes, 1979.

Herneck, Friedrich. "Zu einem Brief Albert Einsteins an Ernst Mach." *PB, 15* (1959), 563–64.

―――. "Zum Briefwechsel Albert Einsteins mit Ernst Mach." *Forschungen und Fortschritte, 37* (1963), 239–43.

―――. *Wissenschaftsgeschichte: Vorträge und Abhandlungen.* Berlin: Akademie-Verlag, 1984.

Hickson, J. W. A. "Recent attacks on causal knowledge." *Philosophical review, 47* (1938), 595–606.

Hoffmann, Banesh. *Albert Einstein, creator and rebel.* New York: Viking, 1972.

Hoffmann, Dieter. "Johannes Stark: eine Persönlichkeit im Spannungsfeld von wissenschaftlicher Forschung und faschistischer Ideologie." Humboldt-Universität, Berlin. Sektion Marxistisch-Leninistische Philosophie. *Wissenschaft und Persönlichkeit.* Berlin: Humboldt-Universität, 1982. Pp. 90–101. (Philosophie und Naturwissenschaft in Vergangenheit und Gegenwart, 22.)

—————. "Max Planck und die Popularisierung naturwissenschaftlicher Erkenntnisse." Urania. Gesellschaft zur Verbreitung wissenschaftlicher Kenntnisse. *Zur Geschichte der Verbreitung wissenschaftlicher Kenntnisse unter den Volksmassen im kapitalistischen Deutschland.* Berlin: Urania, 1982. Pp. 19–28.

—————. "Die Physik an der Berliner Universität in der ersten Hälfte unseres Jahrhunderts. Zur personellen und institutionellen Entwicklung sowie wichtige Beziehungen mit anderen Institutionen physikalischer Forschung in Berlin." AW. ITGOW. *Kolloquien, 35* (1984), 5–30.

—————. "Max Planck als akademischer Lehrer." Ibid., 55–71.

Hoffmann, Peter. *The history of the German resistance, 1933–1945.* Trans. Richard Barry. Cambridge: MIT Press, 1977.

Hoffmeister, Johannes. *Wörterbuch der philosophischen Begriffe.* Leipzig: F. Meiner, 1944.

Holton, Gerald. *Thematic origins of scientific thought. Kepler to Einstein.* Cambridge: Harvard University Press, 1973.

Inge, William Ralph. *God and the astronomers.* London and New York: Longmans Green, 1933.

Institut International de Physique Solvay. *La théorie du rayonnement et les quanta. Rapports et discussions de la réunion tenue à Bruxelles du 30 octobre au 3 novembre 1911.* Paris: Gauthiers-Villars, 1912.

Irving, David. *The German atomic bomb. The history of nuclear research in Nazi Germany.* New York: Simon and Schuster, 1967.

Jánossy, L. "Plancks philosophische Ansichten in der Physik." In Kockel et al., eds., *Max-Planck-Festschrift* (1959), 389–407.

Jensen, Paul. "Kausalität, Biologie, und Psychologie." *Erkenntnis*, 4 (1934), 165–214.

Jordan, Pascual. "Über den positivistischen Begriff der Wirklichkeit." *Nwn*, 22 (1934), 485–90.

———. *Anschauliche Quantentheorie. Eine Einführung in die moderne Auffassung der Quantenerscheinungen*. Berlin: Springer, 1936.

Kaiser-Wilhelm-Gesellschaft. *Liste der Mitglieder*. Berlin: Reichsdruckerei, 1917.

———. "Tätigkeitsbericht." *Nwn*, 20 (1932), 428–61, *Nwn*, 23 (1935), 321–67, and *Nwn*, 24 (1936), 19–48.

Kangro, Hans. *Vorgeschichte des Planckschen Strahlungsgesetzes*. Wiesbaden: Franz Steiner, 1970.

Keussen, Rudolf. *Die Willensfreiheit als religiöses und philosophisches Grundproblem*. Freiburg/Br.: Willibrordbuchhandlung, 1935.

Kirsten, Christa. "Kommentar [zu Plancks Selbstdarstellung]." AW. *Aus der Arbeit von Plenum und Klasse, 8:4* (1983), 4–11.

Kirsten, Christa, and Hans-Günther Körber. *Physiker über Physiker*. 2 vols. Berlin: Akademie-Verlag, 1975–79. (Vol. 1, *Wahlvorschläge zur Aufnahme von Physikern in die Berliner Akademie 1870 bis 1930*; vol. 2, *Antrittsreden . . . , Gedächtnisreden*.)

Kirsten, Christa, and Hans-Jürgen Treder. *Albert Einstein in Berlin, 1913–1933*. 2 vols. Berlin: Akademie-Verlag, 1979.

Klein, Martin J. "Thermodynamics and quanta in Planck's work." *Physics today, 19:11* (1966), 23–32.

———. *Paul Ehrenfest*. Amsterdam: North Holland, 1970.

Kleinert, Andreas. "Eine Stellungnahme Plancks zur Frage des Frauenstudiums." *PB, 34* (1978), 31–33.

————. "Lenard, Stark, und die Kaiser-Wilhelm-Gesellschaft." *PB, 36:2* (1980), 35–38, 43.

Kleinpeter, Hans. "Zur Begriffsbestimmung des Phänomenalismus." *VWPS, 36* (1912), 1–18.

Klohr, Olof. "Max Planck—Naturwissenschaft—Religion." Halle Universität. *Wissenschaftliche Zeitschrift, Math.-Naturw. Reihe, 6:2* (1957), 293–99.

Kockel, B., W. Macke, and A. Papapetrou, eds. *Max-Planck-Festschrift 1959.* Berlin: Deutscher Verlag der Wissenschaften, 1959.

Körber, Hans-Günther, ed. *Aus dem wissenschaftlichen Briefwechsel Wilhelm Ostwalds.* Berlin: Akademie-Verlag, 1961–.

Konen, Heinrich. "Zur Frage des Unterrichtes der Studierenden der Physik an den Universitäten." *ZPCU, 20* (1907), 231–34.

Krafft, Fritz. *Im Schatten der Sensation. Leben und Wirken von Fritz Strassmann.* Weinheim: Verlag Chemie, 1981.

Kropp, Gerhard. "Die philosophischen Gedanken Max Plancks." *Zeitschrift für philosophische Forschung, 6:3* (1952), 434–58.

Kuhn, T. S. *Black-body theory and the quantum discontinuity, 1894–1912.* New York and Oxford: Oxford University Press, 1978.

Künzel, Friedrich. "Max Planck als Sekretär der Berliner Akademie der Wissenschaften in den Jahren 1912 bis 1938." AW. ITGOW. *Kolloquien, 35* (1984), 73–84.

Langevin, André. *Paul Langevin, mon père.* Paris: Editeurs français réunis, 1971.

Laue, Max von. "Steiner und die Naturwissenschaft." *Deutsche Revue, 47* (1922), 41–49. (Also in Laue, *Gesammelte Schriften, 3,* 48–56.)

————. "Zu der Erörterung über Kausalität." *Nwn, 20* (1932), 915–16.

————. "Ansprache bei der Eröffnung der Physikertagung in Würzburg." *Physikalische Zeitschrift, 34* (1933), 889–90. (Also in Laue, *Gesammelte Schriften, 3,* 61–62.)

————. "Materie und Raumerfüllung." *Scientia*, *54* (1933), 402–12, and Suppl., 182–91.

————. "Nachruf auf Fritz Haber." *Nwn*, *22* (1934), 97.

————. "Über Heisenbergs Ungenauigkeitsbeziehungen und ihre erkenntnistheoretische Bedeutung." *Nwn*, *22* (1934), 439–40.

————. "Max Planck." *Nwn*, *35* (1948), 1–7.

————. "Zu Max Plancks 100. Geburtstage." *Nwn*, *45* (1958), 223.

————. *Gesammelte Schriften und Vorträge*. 3 vols. Braunschweig: Vieweg, 1961.

Lenard, Philipp. *Quantitatives über Kathodenstrahlen aller Geschwindigkeiten*. Heidelberg, 1918. (Akademie der Wissenschaften, Heidelberg. Math.-Naturw. Klasse. *Abhandlungen*, 5.)

————. *Deutsche Physik*. 4 vols. Munich: J. F. Lenmanns, 1936–37.

————. "Erinnerungen eines Naturforschers, der Kaiserreich, Judenherrschaft und Hitlerzeit erlebt hat." Typescript autobiography, AHQP.

Lenz, Max. *Geschichte der königlichen Friedrich-Wilhelms-Universität zu Berlin*. 4 vols. Halle: Buchhandlung des Waisenhauses, 1910–18.

Lexikon für Theologie und Kirche. Freiburg/Br.: Herder, 1938.

Lindner, Helmut. "'Deutsche' und 'gegentypische' Mathematik." Mehrtens and Richter, eds., *Naturwissenschaft, Technik, und NS-Ideologie* (1980), 88–115.

Lowood, Henry. *Max Planck. A bibliography of his non-technical writings*. Berkeley: Office for History of Science and Technology, University of California, 1977.

Ludloff, Rudolf. "Der Aufenthalt deutscher Hochschullehrer in Moskau und Leningrad 1925." Jena. Universität. *Wissenschaftliche Zeitschrift, Gesellschafts- und Sprachwissenschaftliche Reihe*, 1956/7:6, 709–21.

Mach, Ernst. "Die Leitgedanken meiner naturwissenschaftlichen Erkenntnislehre und ihre Aufnahme durch die Zeitgenossen." *Physikalische Zeitschrift, 11* (1910), 599–606.

Martens, Erika. *Zum Beispiel "Das Reich." Zur Phänomenologie der Presse im totalitären Regime.* Cologne: Verlag Wissenschaft und Politik, 1972.

Max-Planck-Gesellschaft. *50 Jahre Kaiser-Wilhelm-Gesellschaft und Max-Planck-Gesellschaft zur Förderung der Wissenschaften.* Göttingen: MPG, 1961.

"Max Planck zum Gedächtnis." *PB, 4* (1948), 133–80. (Contributions by Alfred Bertholet, Louis de Broglie, Felix Ehrenhaft, Heinrich von Ficker, Axel von Harnack, Friedrich Hoffmann, Pascual Jordan, Ernst Lamla, J. R. Partington, Wilhelm Westphal, and Agnes von Zahn-Harnack.)

McCormmach, Russell. *Night thoughts of a classical physicist.* Cambridge: Harvard University Press, 1982.

Mehrtens, Herbert, and Steffen Richter, eds. *Naturwissenschaft, Technik und NS-Ideologie: Beiträge zur Wissenschaftsgeschichte des Dritten Reiches.* Frankfurt/Main: Suhrkamp, 1980.

Meissner, Walther. "Max Planck, the man and his work." *Science, 113* (1951), 75–81.

Meitner, Lise. "Max Planck als Mensch." *Nwn, 45* (1958), 406–8.

―――. "Looking back." *Bulletin of the atomic scientists, 20:11* (1964), 2–7.

Merten, Bernhard. "Max Planck: Das Weltbild der neuen Physik." *Nwn, 25* (1937), 253–54.

Müller, Friedrich von. "Eröffnungsrede." GDNA. *Verhandlungen, 86* (1920), 15–24.

Müller, Wilhelm. "Jüdischer Geist in der Physik." *Zeitschrift für die gesamte Naturwissenschaft, 5* (1939), 162–75.

―――. "Die Lage der theoretischen Physik an den Universitäten." Ibid., *6* (1940), 282–98.

Nagel, Bengt. "The discussion concerning the Nobel prize for

Max Planck." In Bernhard et al., *Science, technology, and society* (1982), 352–76.

Nathan, Otto, and Heinz Norden, eds. *Einstein on peace*. New York: Avenel, 1981.

Noack, Karl. "Die Vorbildung der Physiklehrer." *ZPCU*, *20* (1907), 147–53.

O'Flaherty, James C. "Max Planck and Adolf Hitler." American Association of University Professors. *Bulletin*, *42* (1956), 437–44.

———. "A humanist looks at Max Planck." *American scientist*, *47* (1959), 68–79.

Ostwald, Wilhelm. "Max Planck: Die Einheit des physikalischen Weltbildes." *Annalen der Naturphilosophie*, *10* (1910), 105–6.

———. *Lebenslinien. Eine Selbstbiographie*. 3 vols. Berlin: Klasing, 1926–27.

Pachaly, Erhard. "Die Beziehungen der Notgemeinschaft der deutschen Wissenschaft zur sowjetschen Wissenschaft." In Bielfeldt et al., eds., *Deutschland, Sowjetunion* (1966), 129–37.

Pauli, Wolfgang. *Wissenschaftlicher Briefwechsel*. Ed. A. Hermann, K. von Mayenn, and V. F. Weisskopf. New York, Heidelberg, and Berlin: Springer, 1979–.

Peter, Karl Heinrich. *Spiegelbild einer Verschwörung. Die Kaltenbrunner-Berichte an Bormann und Hitler über das Attentat vom 20. Juli 1944*. Stuttgart: Seewald, 1961.

Petzold, Joseph. "Die vitalistische Reaktion auf die Unzulänglichkeit der mechanischen Naturansicht." *Zeitschrift für allgemeine Physiologie*, *10* (1910), 69–119.

Picard, Emile. "Sciences." Paris. Exposition Universelle Internationale de 1900. Jury International. *Rapports. Introduction générale*, *2*, vii–ix, 1–114. Paris: Imprimerie nationale, 1903.

Pihl, Gunnar T. *Germany. The last phase*. New York: Knopf, 1944.

Planck, Max. *Grundriss der allgemeinen Thermochemie. Mit ei-*

nem Anhang: Der Kern des zweiten Hauptsatzes der Wärmetheorie. Breslau: Eduard Trewendt, 1893.

———. *Vorlesungen über Thermodynamik.* 2nd ed. Leipzig: von Veit, 1905. (1st ed., 1897.)

———. *Das Prinzip der Erhaltung der Energie. Von der Philosophischen Fakultät Göttingen preisgekrönt.* 2nd ed. Leipzig and Berlin: Teubner, 1908. (1st ed., 1887.)

———. *Acht Vorlesungen über theoretische Physik gehalten an der Columbia University in the City of New York im Frühjahr 1909.* Leipzig: Hirzel, 1910.

———. "Zur Machschen Theorie der physikalischen Erkenntnis. Eine Erwiderung." *Physikalische Zeitschrift, 11* (1910), 1186–90.

———. "[Ansprache]." *Berliner akademische Nachrichten, 8* (1913–14), 26–27.

———. *Das Weltbild der neuen Physik.* Leipzig: Barth, 1929.

———. *Einführung in die Theorie der Wärme.* Leipzig: Hirzel, 1930.

———. "Dem Andenken an W. Wien." In Wien, *Aus dem Leben* (1930), 139–41.

———. "Neue Erkenntnisse der Physik." *Die Woche,* 1931, 1419–20.

———. "Ein Blick in das Universum. Die Stellung und Bedeutung der heutigen Physik." *Ernte, 13:7* (1932), 31–33.

———. "Überproduktion von wissenschaftlichen Ideen." *Berliner Tageblatt,* Sonntags-Ausgabe (1. Beiblatt), 18 June 1933.

———. "Arnold Sommerfeld zum siebzigsten Geburtstag." *Nwn, 26* (1938), 777–79.

———. "Max von Laue. Zum 9. Oktober 1939." *Nwn, 27* (1939), 665–66.

———. "Naturwissenschaft und reale Aussenwelt." *Nwn, 28* (1940), 778–79.

_____. "Warum kann Wissenschaft nicht populär sein?" *Das Reich* (Berlin), no. 32, 9 August 1942.

_____. "Selbstdarstellung (1942)." AW, Berlin. *Aus der Arbeit von Plenum und Klasse, 8:4* (1983), 4–11.

_____. *Wege zur physikalischen Erkenntnis.* 3rd ed. Leipzig: Hirzel, 1944.

_____. "Mein Besuch bei Adolf Hitler." *PB, 3* (1947), 143.

_____. *Erinnerungen.* Ed. W. Keiper. Berlin: Keiperverlag, 1948.

_____. *Physikalische Abhandlungen und Vorträge.* 3 vols. Braunschweig: Vieweg, 1958.

_____. *The new science.* New York: Meridian, 1959.

Poliakov, Leon, and Josef Wulf, eds. *Das Dritte Reich und seine Denker.* Berlin: Arani, 1959.

Poske, Friedrich. "Die Einrichtungen für den naturwissenschaftlichen Unterricht an den höheren Schulen Preussens." *ZPCU, 20* (1907), 401–4.

Przibram, K., ed. *Schrödinger, Planck, Einstein, Lorentz. Briefe zur Wellenmechanik.* Vienna: Springer, 1963.

Pyenson, Lewis. *Neohumanism and the persistence of pure mathematics in Wilhelmian Germany.* Philadelphia: American Philosophical Society, 1983.

_____. *The young Einstein. The advent of relativity.* Bristol and Boston: Adam Hilger, 1985.

Ramsauer, Carl. *Physik-Technik-Pädagogik. Erfahrungen und Erinnerungen.* Karlsruhe: G. Braun, 1949.

Realencyclopädie für protestantische Theologie und Kirche. 3rd ed. Ed. Albert Hauck. Leipzig: J. C. Hinrichs'sche Buchhandlung, 1896–1913.

Reingold, Nathan, and Ida Reingold, eds. *Science in America. A documentary history, 1900–1939.* Chicago: Chicago University Press, 1981.

Reishaus-Etzold, Heike. "Die Einflussnahme der Chemiemonopole auf die 'Kaiser-Wilhelm-Gesellschaft zur Förderung der

Wissenschaften E.V.' während der Weimarer Republik." *JW*, 1973:1, 37–61.

Richter, Steffen. *Forschungsförderung in Deutschland, 1920–1936*. Düsseldorf: VDI, 1972.

———. "Der Kampf innerhalb der Physik in Deutschland nach dem Ersten Weltkrieg." *Sudhoffs Archiv*, 57 (1973), 195–207.

———. "Die 'Deutsche Physik.'" In Mehrtens and Richter, *Naturwissenschaft, Technik und NS-Ideologie* (1980), 116–41.

Ringer, Fritz. *The decline of the German mandarins*. Cambridge: Harvard University Press, 1969.

Runge, Iris. *Carl Runge und sein wissenschaftliches Werk*. Göttingen: Vandenhoeck and Ruprecht, 1949.

Salomon, Felix. *Die deutschen Parteiprogramme. Heft 3. Von der Revolution bis zum neuen Reichstag, 1918–1920*. 3rd ed. Leipzig and Berlin: Teubner, 1920.

Schäfer, Dietrich. "Festrede." Berlin. Universität. *Feier . . . zur Erinnerung an die Erhebung der deutschen Nation im Jahre 1813*. Berlin: Universitäts-Buchdruckerei, 1913. Pp. 5–29.

Schenck, Rudolf. "Arbeitsgemeinschaft und Gemeinschaftsarbeit in Naturwissenschaften und Technik." In Abb, ed., *Aus fünfzig Jahren* (1930), 286–99.

Scherzer, Otto. "Physik im totalitären Staat." In Andreas Flitner, ed. *Deutsches Geistesleben und Nationalsozialismus*. Tübingen: Rainer Wunderlich, 1965. Pp. 47–58.

Schlicker, Wolfgang. "Forschung und Gesellschaft—Vergesellschaftung der Forschung. Zur Stellung der Notgemeinschaft der deutschen Wissenschaft in der bürgerlichen Forschungspolitik der Weimarer Republik." *NTM. Schriftenreihe für Geschichte der Naturwissenschaft, Technik und Medizin*, 12:1 (1975), 45–55.

———. "Zu Max Plancks Bedeutung für die Leitung der Wissenschaft und Organisation der Forschung unter besonderer Berücksichtigung seines Wirkens in der Weimarer Republik." *JW*, 1975:2, 161–85.

Schmidt-Ott, Friedrich. *Erlebtes und Erstrebtes, 1860–1950.* Wiesbaden: Franz Steiner, 1952.

Scholder, Klaus, ed. *Die Mittwochs-Gesellschaft. Protokolle aus dem geistigen Deutschland 1932 bis 1944.* Berlin: Severin and Siedler, [1982].

Schreber, Karl. "Zum Unterricht in der Experimentalphysik auf den Universitäten. *ZPCU, 19* (1906), 213–15.

Schreiber, Reinhold. "Auslandsbeziehungen der deutschen Wissenschaft." In Abb, ed., *Aus fünfzig Jahren* (1930), 9–21.

Schroeder-Gudehus, Brigitte. *Deutsche Wissenschaft und internationale Zusammenarbeit, 1914–1928.* Geneva: Dumaret and Golay, 1966.

———. "The argument for the self-government and public support of science in Weimar Germany." *Minerva, 10* (1972), 537–70.

———. *Les scientifiques et la paix.* Montreal: Presses de l'Université, 1978.

Schrödinger, Erwin. "Was ist ein Naturgesetz?" *Nwn, 17* (1929), 9–11. (Text of 9 Dec. 1922.)

Schwabe, Klaus. *Wissenschaft und Kriegsmoral. Die deutschen Hochschullehrer und die politischen Grundfragen des Ersten Weltkrieges.* Göttingen: Musterschmidt, 1969.

[*Science.*] "The scientific situation in Germany." *Science, 77* (1933), 528–29.

Seelig, Carl. *Albert Einstein und die Schweiz.* Zurich: Europa-Verlag, 1952.

———. *Helle Zeit—Dunkle Zeit.* Zurich: Europa-Verlag, 1956.

Shea, W. R., ed. *Otto Hahn and the rise of nuclear physics.* Dordrecht and Boston: Reidel, 1983.

Sommerfeld, Arnold. "[Ernst Mach.]" Akademie der Wissenschaften, Munich. *Jahrbuch,* 1917, 58–67.

———. "Max Planck zum sechzigsten Geburtstage." *Nwn, 5* (1918), 195–202.

———. "Die Entwicklung der Physik in Deutschland seit Heinrich Hertz." *Deutsche Revue*, *43:3* (1918), 122–32.

———. "Über die Anfänge der Quantentheorie von mehreren Freiheitsgraden." *Nwn*, *17* (1929), 481–83.

Stamm, Thomas. *Zwischen Staat und Selbstverwaltung. Die deutsche Forschung im Wiederaufbau 1945–1965.* Cologne: Verlag Wissenschaft und Politik, 1981.

Stark, Johannes. *Die Prinzipien der Atomdynamik.* 3 vols. Leipzig: Hirzel, 1910–15.

———. "The attitude of the German government towards science." *Nature*, *133* (21 Apr. 1934), 614.

———. *Adolf Hitler und die deutsche Forschung.* Berlin: Pass and Garleb, [1934].

———. "Einstein, Heisenberg und Schrödinger." *Nationalsozialistische Monatshefte*, no. 71, 1936. In Poliakov and Wulf, eds., *Das Dritte Reich* (1959), 300–308.

———. "The pragmatic and the dogmatic spirit in physics." *Nature*, *141* (30 Apr. 1938), 770–72.

Stern, Leo, ed. *Die Berliner Akademie der Wissenschaften in der Zeit des Imperialismus.* 3 vols. Berlin: Akademie-Verlag, 1975–79. (Vol. 1, 1900–1917, by Conrad Grau; vol. 2, 1917–1933, by Wolfgang Schlicker; vol. 3, 1933–45, by Conrad Grau, Wolfgang Schlicker, and Liane Zeil.)

Tauber, Gerald E., ed. *Albert Einstein's theory of general relativity.* New York: Crown, 1979.

Thiele, Joachim. "Ein zeitgenössisches Urteil über die Kontroverse zwischen Max Planck und Ernst Mach." *Centaurus*, *13* (1968), 85–90.

Unsöld, Albrecht. *Physik und Historie. Rede und Ansprachen anlässlich der Enthüllung des Kieler Max-Planck-Denkmals am 23. April 1958.* Kiel: Ferdinand Hirt, 1958.

Vogel, Heinrich. *Zum philosophischen Wirken Max Plancks. Seine Kritik am Postivismus.* Berlin: Akademie-Verlag, 1961.

———. "Die Stellung Max Plancks zur Sowjetunion und ihren Wissenschaftlern, dargestellt am Beispiel seiner Reise im

Jahre 1925." In Bielfeldt et al., eds., *Deutschland, Sowjet-union* (1966), 466–72.

Wehberg, Hans. *Wider den Aufruf der 93! Das Ergebnis einer Rundfrage an die 93 Intellektuellen über die Kriegsschuld.* Berlin: Verlag für Politik und Geschichte, 1920.

Wendel, Günter. "Der Kampf deutscher Naturwissenschaftler gegen die Vorbereitung eines Atomkrieges." In Harig and Mette, eds., *Naturwissenschaft-Tradition-Fortschritt* (1963), 327–42.

———. *Die Kaiser-Wilhelm-Gesellschaft, 1911–1914. Zur Anatomie einer imperialistischen Forschungsgesellschaft.* Berlin: Akademie-Verlag, 1975.

Westphal, Wilhelm H. "Max Planck als Mensch." *Nwn, 45* (1958), 234–36.

Weyl, Hermann. "Die Relativitätstheorie auf die Naturforscherversammlung in Bad Nauheim." Deutsche Mathematiker-Vereinigung. *Jahresbericht, 31* (1922), 51–63.

Wheaton, Bruce R. "Inventory of sources for history of twentieth-century physics (ISHTCP)," *Isis, 75* (1984), 153–57.

Wiedemann, Eilhard. "Über den Hochschulunterricht für künftige Lehrer der Physik." *ZPCU, 19* (1906), 265–71.

Wien, Wilhelm. *Ziele und Methoden der theoretischen Physik.* Würzburg, 1914. (Speech at the celebration of the 232nd anniversary of the University of Würzburg.)

———. *Aus dem Leben und Wirken eines Physikers.* Ed. Karl Wien. Leipzig: Barth, 1930.

Willstätter, Richard. *Aus meinem Leben. Von Arbeit, Musse, und Freunden.* Munich: Verlag Chemie, 1949.

Zahn-Harnack, Agnes von. *Adolf von Harnack.* 2nd ed. Berlin: De Gruyter, 1951.

INDEX

243